International Nuclear Science Technology and Engineering Conference (iNuSTEC2024)

IOP Conference Series: Materials Science and Engineering
Volume 1326

Selangor, Malaysia
14-16 November 2024

ISBN: 979-8-3313-2005-8
ISSN: 1757-8981

Printed from e-media with permission by:

Curran Associates, Inc.
57 Morehouse Lane
Red Hook, NY 12571

Some format issues inherent in the e-media version may also appear in this print version.

Institute of Physics
Dirac House, Temple Back
Bristol BS1 6BE UK

Phone: 44 1 17 929 7481
Fax: 44 1 17 920 0979

techtracking@iop.org

Additional copies of this publication are available from:

Curran Associates, Inc.
57 Morehouse Lane
Red Hook, NY 12571 USA
Phone: 845-758-0400
Fax: 845-758-2633
Email: curran@proceedings.com
Web: www.proceedings.com

TABLE OF CONTENTS

IOP Conf. Series: Materials Science and Engineering **1326** (2025) 011001 doi:10.1088/1757-899X/1326/1/011001

IOP Conference Series: Materials Science and Engineering

International Nuclear Science, Technology and Engineering Conference 2024 (iNuSTEC2024)

"Nuclear Science and Technology For Socio-economic Development "

Bangi, Selangor, Malaysia
14 -16 October 2024

Editors

Abdul Aziz Mohamed, Malaysian Nuclear Society
Faridah Mohamad Idris, University Tenaga Nasional
Julie Andrianny Murshidi, Malaysian Nuclear Agency
Azraf Azman, Malaysian Nuclear Agency
Nazaratul Ashifa Abdullah Salim, Malaysian Nuclear Agency
Mazleha Maskin, Malaysian Nuclear Agency
Mohd Idzat Idris, Universiti Kebangsaan Malaysia
Asyraf Arif Abu Bakar, Malaysian Nuclear Agency
Khairul Nadzmi Jamaluddin, Universiti Teknologi Malaysia
Anwar Abdul Rahman, Malaysian Nuclear Agency
Muhammad Amir Asyraf Ramli, Arey Solutions Sdn Bhd
Zaredah Hashim, Malaysian Nuclear Agency

All papers have been peer-reviewed

Supporting/Sponsoring Organisations

Ministry of Education Malaysia; Ministry of Science, Technology and Innovation Malaysia; Malaysian Nuclear Agency; ASME Malaysia Section; Malaysian Welding and Joining Society; Reactor Interest Group; Asia-Oceania Neutron Scattering Association, Japan Atomic Energy Society; Institute of Nuclear Energy, Universiti Tenaga Nasional

IOP Conf. Series: Materials Science and Engineering **1326** (2025) 011001 doi:10.1088/1757-899X/1326/1/011001

MESSAGE FROM:
PRESIDENT, MALAYSIAN NUCLEAR SOCIETY (MNS)

I am really pleased to congratulate all keynote speakers, sponsors, presenters, and participants for the contribution and participating in the International Nuclear Science, Technology and Engineering Conference (iNuSTEC2024). Ideally, the iNuSTEC2024 is organized to enhance the importance of nuclear science, technology, and engineering in developing socio-economic in Malaysia. Successful conferences from previous years, at both national and international levels, have brought this event to be more meaningful to strengthen nuclear development for global growth under challenging situation

With the support from the American Society of Mechanical Engineers (ASME) Malaysia Section, and Malaysian Welding and Joining Society (MWJS), the MNS is indeed honored to co-organise this event with organizations such as Nuclear Malaysia Agency, UNITEN, UTM, UKM, UM, UiTM, USIM and UTHM. The affiliation of the Institute of Nuclear Energy of UNITEN and advanced Nuclear Engineering Research Group of UTM and Institute of Nuclear Engineers (INE) of MNS, to make iNuSTEC2024 a success is a good testament that nuclear still remain relevant in the heart of the young generation to move towards greater strength in nuclear fields. In the future, MNS members shall strive to reach out to more diverse population in order to make nuclear ecosystem more conducive and stronger.

This conference had been supported by experienced and young researchers that can give a platform for them to discuss and debate on the latest development in all areas of nuclear technology applications for friendly purposes at the international level. This would spur and catalyse further research and development leading to the sustainability and improvement of the quality of life without compromising the quality of the environment in Malaysia and the world.

MNS is a unique non-governmental organisation that is formed and headed by individuals that are aligned to the beneficial and peaceful usage of nuclear science, engineering, and technology which has been incorporated under the Malaysian Society Act. The MNS has been actively promoting the information, advancement, popularization, and understanding of nuclear the public in Malaysia and this conference is one of the ways of realizing such aim. This conference is also served as a public acceptance program to introduce the benefits and advantages of the use of nuclear science, engineering, and technology to the public on a large scale. On top of that MNS also published twice a year on regular basis the Journal of Nuclear and Related Technologies (JNRT) and Buletin Nuklear Malaysia (BNM) for researchers to share their research findings with the community. We also publish iNuSTEC proceedings on yearly basis.

I would like to express my thanks to all parties involved in making iNuSTEC2024, Nuclear Youth Competition 2024 and post workshop/seminar on Small Modular Reactor (SMR) a success. I wish to express my heartful thanks to all speakers for your willingness to share your experience with the participants and presenters for the development of the nuclear field. The contribution is not only coming from the Malaysian scientific community but also distinguished guest speakers of iNuSTEC2024. to all the personnel involved for their unwavering efforts to make this event a memorable event. May all these efforts bring benefits to all of us and I congratulate everyone involved in making this conference successful event.

Associate Professor Dr Abdul Aziz Mohamed
President
Malaysian Nuclear Society

IOP Conf. Series: Materials Science and Engineering **1326** (2025) 011001 doi:10.1088/1757-899X/1326/1/011001

CONTENTS

PAPERS

IOP Conf. Series: Materials Science and Engineering **1326** (2025) 011001 doi:10.1088/1757-899X/1326/1/011001

iNuSTEC2024 ORGANISING COMMITTEE

Advisor:
Prof. Dr. Nahrul Khair Alang Md Rashid
Advisor, Malaysian Nuclear Society/ Universiti Teknologi Malaysia

Chairman:
Assoc. Prof. Dr Aziz Mohamed, President, MNS

Deputy Chairman:
Azraf Azman, Deputy President, MNS

Secretary:
Dr Faridah Mohamad Idris, MNS/UNITEN

Deputy Secretary
Dr Mazleha Maskin, MNS/Nuklear Malaysia

Treasurer:
Dr Julie Andrianny, MNS Treasurer

Deputy Treasurer:
Dr Nazaratul Ashifa Abdullah Salim, MNS Treasurer

Advisory Panel
Dato' Seri Prof. Ir. Dr. Noor Azuan Abu Osman, UM
Prof Datuk Dr Shahrin Bin Sahib @ Sahibuddin, UiTM
Prof. Ir. Dr. Khairul Salleh Mohamed Sahari, UNITEN
Prof. Ts. Dr. Sharifudin Md. Shaarani, USIM
Prof. Dato' Gs. Ts. Dr. Mohd Ekhwan Hj. Toriman, UKM
Dr. Rosli Bin Darmawan, Nuklear Malaysia
Datuk Ts. Dr. Ahmad Fauzi Ismail Naib Canselor, UTM
Ts. Dr. Ruzairi Bin Abdul Rahim, UTHM

IOP Conf. Series: Materials Science and Engineering **1326** (2025) 011001 doi:10.1088/1757-899X/1326/1/011001

TECHNICAL COMMITTEE

Prof. Dr Faizal Mohamed, MNS/UKM
Dr Mohd Idzat Bin Idris, MNS/UKM
Dr Khairul Nadzmi Bin Jamaluddin, MNS/UTM
Asyraf Arif Abu Bakar, MNS/Nuklear Malaysia
Muhammad Amir Asyraf bin Ramli, MNS/Arey Solutions
Anwar B. Abdul Rahman, MNS/Nuklear Malaysia
Zaredah Hashim, MNS/Nuklear Malaysia
Dr Norsyahidah binti Mohd Hidzir, UKM
Dr Muhammad Arif Sazali, INE MNS/UTM
Dr Muhammad Syahir Sarkawi, INE MNS/UTM
Dr Nur Syazwani Mohd Ali, INE MNS/UTM
Dr Mohd Syukri Yahya, MNS/Uniten
Dr Hassan Mohaed, MNS/Uniten
Mohd Rizal B. Mamat @ Ibrahim, MNS/Nuklear Malaysia
Dr. Hafizal Yazid, MNS/Nuklear Malaysia
Shalina Bt. Sheik Muhamad, MNS/Nuklear Malaysia
Dr Mohd Hafez bin Mohd Isa, USIM
Prof Hasan Abu Kassim, MNS/UM
Assoc Prof Dr Norhasliza Yusof, MNS/UM
Dr Mohsin Mohd Saies, UTM
Dr Khaidzir Hamzah, MNS/UTM
Prof Dr Zaidi Embong, MNS/UTHM
Puteri Nuraliah Husna Binti Mohd Tajuddin, INE MNS/Nuklear Malaysia
Khairul Anwar Bapujee, INE MNS/Nuklear Malaysia
Muhammad Zulhelmi Mahadi, INE MNS/N

IOP Conf. Series: Materials Science and Engineering **1326** (2025) 011001 doi:10.1088/1757-899X/1326/1/011001

NUCLEAR YOUTH COMPETITION 2024

Champion:

1. Safiyah Binti Iwan Safian, Universiti Kebangsaan Malaysia (UKM)
2. Nur Ain Sofiyah Binti Mohamed, Universiti Sains Islam Malaysia (USIM)
3. Nur 'Alea Najeeha Binti Zazuzila, Universiti Sains Islam Malaysia (USIM)
4. Henry Kam Meng Tung, Universiti Teknologi Malaysia (UTM)

2nd Place:

1. Safwa Azalea Binti Shahrulinizam, Universiti Kebangsaan Malaysia (UKM)
2. Siti Salwani Binti Hasbullah, Universiti Sains Islam Malaysia (USIM)
3. Nurin Ain Najwa Binti Rosmaizi, Universiti Sains Islam Malaysia (USIM)
4. Isma Rusyaidi Bin Iskandar, Universiti Sains Islam Malaysia (USIM)
5. Ho E-Sian, Universiti Teknologi Malaysia (UTM)

3rd Place:

1. Sharifah Wajihah Binti Long Mohd Nor Azizi, Universiti Kebangsaan Malaysia (UKM)
2. Izzatul Faqihah Binti Mohd Azahar, Universiti Sains Islam Malaysia (USIM)
3. Adlina Batrisyia Binti Muhd Nizam, Universiti Sains Islam Malaysia (USIM)
4. Amni Batrisyia Binti Ayub, Universiti Sains Islam Malaysia (USIM)
5. Tuan Fatin Faqihah Binti Tuan Arif Sahibu Azemi, Universiti Teknologi Malaysia (UTM)

IOP Conf. Series: Materials Science and Engineering **1326** (2025) 011001 doi:10.1088/1757-899X/1326/1/011001

iNuSTEC2024 POSTER COMPETITION

Champion:

Revolutionizing Cancer Treatment: Harnessing 68Ga-NOTA-RITUXIMAB in Radioimmunotherapy (RIT)

Siti Aminah Muhamad, Nadhirah Razanah Shahrol, Nurul Nazeerah Juarimi, Saifullizan Mohamad, Wan Hamirul Bahrin Wan Kamal

2nd Place:

Total Phosphate Removal From Xenotime Mineral via Alkaline Fusion

Roshasnorlyza Hazan, Nurhanisah Asri, Muhammad Nu'aim, Nur Aqilah Sapiee, Khaironie Mohamed Takip, Wilfred Paulus, Norhazirah Azhar and Jacqueline Kones

3rd Place:

Production and Quality Assessment of Radiopharmaceutical-Grade Zoledronic Acid Kit for SPECT Imaging

Ng Yen, Manisah Saedon, Muhammad Fathurrahman Faizul Akhmar, Muhammad Ridzuan Rosdi and Wan Hamirul Bahrin Wan Kamal

IOP Conf. Series: Materials Science and Engineering **1326** (2025) 011001 doi:10.1088/1757-899X/1326/1/011001

ACKNOWLEDGEMENT

iNuSTEC2024 is organized by Malaysian Nuclear Society (MNS)

In Collaboration with

Malaysian Nuclear Agency (Nuclear Malaysia)
Universiti Teknologi Malaysia (UTM)
Universiti Tenaga Nasional (UNITEN)
Universiti Kebangsaan Malaysia (UKM)
Universiti Teknologi MARA (UiTM)
Universiti Islam Antarabangsa Malaysia (IIUM)
Universiti Malaya (UM)
Universiti Sains Islam Malaysia (USIM)
Universiti Tun Hussien Onn Malaysia (UTHM)

Co-organized by
National Centre for Particle Physics (NCPP)
Institute of Nuclear Engineers (INE-MNS)
Kelab Sains Nuklear (KESAN)
Nuclear Engineering Student Society (NESS)

Supported by

ASME Malaysia Section
Malaysian Welding and Joining Society

IOP Conf. Series: Materials Science and Engineering **1326** (2025) 011001 doi:10.1088/1757-899X/1326/1/011001

With Compliments

Sponsors

CHEMBIO TECHNOLOGY SDN BHD

SYAIZA ENGINEERING SDN BHD

IOP Conf. Series: Materials Science and Engineering **1326** (2025) 011001 doi:10.1088/1757-899X/1326/1/011001

MNS COUNCIL MEMBERS 2023/2025

Advisor / Past President: Prof. Dr. Nahrul Khair Bin Alang Md Rashid (*UTM*)
President Dr. Abdul Aziz Bin Mohamed (*MNS*)
Vice President Azraf Bin Azman (*Nuklear Malaysia*)
Secretary Prof (Adj) Dr. Faridah Bt Mohamad Idris (*UNITEN/MNS*)
Deputy Secretary Dr Mazleha Maskin (*Nuklear Malaysia*)
Treasurer Dr. Julie Andrianny Bt. Murshidi (*Nuklear Malaysia*)
Deputy Treasurer Dr Nazaratul Ashifa Binti Abdullah Salim (*Nuklear Malaysia*)

Members
Dr Mohd Idzat Idris (*UKM*)
Asyraf Arif Abu Bakar (*Nuklear Malaysia*)
Dr Khairul Nadzmi Bin Jamaluddin (UTM)
Anwar Abdul Rahman (*Nuklear Malaysia*)
Dr. Mohd Syukri Yahya (*UNITEN*)

Auditors
Muhammad Amir Asyraf bin Ramli (*Arey Solutions*)
Zaredah Hashim (*Nuklear Malaysia*)

MNS Chapters
RadTech (*Radiation Tech*)
XApp (*X-ray App*)
NuEnergy (*Nuclear Energy*)
Neutron Photon (*Neutron- Photon*)
Student Chapter (*UKM, UiTM, UNITEN, UTM - NESS*)
Women in Nuclear, Malaysia (*WiN Malaysia*)
Institute of Nuclear Engineers (*INE MNS*)
Malaysian Nuclear Professional Chapter (*MNPC MNS*)

MNPC-MNS COMMITTEE MEMBERS 2022/2023
Advisor / President MNS: Dr. Abdul Aziz Bin Mohamed
President: Ts. Razali Bin Harun (*REM*)
Vice President: Khairul Izzat Bin Abdul Aziz (*Reltech Lab*)
Secretary: Mohd Taha Bin Hj. Suja' (*REM*)

INE MNS COMMITTEE MEMBERS 2022/2023
President: Asyraf Arif Bin Abu Bakar (Nuklear Malaysia)
Vice President: Khairul Anwar Bin Bapujee (Nuklear Malaysia)
Secretary: Puteri Nuraliah Husna Binti Mohd Tajuddin (Nuklear Malaysia)
Committee: Dr Muhammad Syahir Sarkawi (UTM),
Muhammad Zulhelmi Bin Mahadi (Nuklear Malaysia)

Visit us at http://www.nuklearmalaysia.org

IOP Conf. Series: Materials Science and Engineering **1326** (2025) 011002 doi:10.1088/1757-899X/1326/1/011002

Peer Review Statement

All papers published in this volume have been reviewed through processes administered by the Editors. Reviews were conducted by expert referees to the professional and scientific standards expected of a proceedings journal published by IOP Publishing.

- **Type of peer review:** Single Anonymous
- **Conference submission management system:** Morressier
- **Number of submissions received:** 16
- **Number of submissions sent for review:** 16
- **Number of submissions accepted:** 16
- **Acceptance Rate (Submissions Accepted / Submissions Received × 100):** 100
- **Average number of reviews per paper:** 2.87
- **Total number of reviewers involved:** 20
- **Contact person for queries:**
  Name: Faridah Mohamad Idris
  Email: faridah246@gmail.com
  Affiliation: Universiti Tenaga Nasional

IOP Conf. Series: Materials Science and Engineering **1326** (2025) 012001 doi:10.1088/1757-899X/1326/1/012001

Measurement of mass attenuation coefficients of several water equivalent materials in comparison to water by using x-ray effective energies

Mohd Fahmi Mohd Yusof[1,2], Betsy Anak Gepat[1], Nurul Shazliana Mohd Daud[1], Che Nazri Che Hussin[1]

[1]School of Health Sciences, Universiti Sains Malaysia, 16150 Kota Bharu, Kelantan
[2]Faculty of Health & Life Science, Management & Science University, 40100 Shah Alam, Selangor

mohdfahmi_mohdyusof@msu.edu.my

Abstract. The mass attenuation coefficient becomes an important parameter to determine the attenuation property of a material towards photons. The study focused on the measurement of mass attenuation coefficients of several phantom materials in comparison to water by using effective energies of kilovoltage X-rays. The mass attenuation coefficient of Perspex®, solid water phantom and paraffin wax regularly used in diagnostic radiology and radiotherapy is measured by using effective energies of 60, 81 and 125 kVp X-rays based on the common diagnostic imaging set ups and compared to water. The results indicated that the mass attenuation coefficients of Perspex®, solid water and paraffin wax are in agreement with water with percentage of discrepancies between 22 and 13%. The percentage differences to water also becoming smaller when higher x-ray energies are used. The results indicated the suitability of the X-ray effective energy method for the measurement of mass attenuation coefficients of materials.

1. Introduction

The mass attenuation coefficients μ/ρ became the important parameter to investigate the attenuation properties of materials towards ionising radiation. It has been used in many previous studies on the fabricated tissue equivalent materials as phantoms (Nair et al., 2017; Tekin et al., 2018; Frimaio et al., 2019; Kargar et al., 2019). Previous studies reported that the XCOM software as simulation method to calculate the attenuation coefficients of elements, compounds and mixture at broad photon energies (Marashdeh et al., 2012; Yusof et al., 2017). This software has been widely used for the determination of mass attenuation coefficients of materials at various photon energy ranges.

The mass attenuation coefficient values of materials can be achieved experimentally with monoenergetic beam. The production of ideal monoenergetic beam itself remains a tough task and researchers always came out with diverse methods in obtaining monoenergetic radiation beam (Frimaio et al., 2019; Mousa et al., 2017). Spectrometer, synchrotrons and parametric x-ray sources are among the methods used in the production of monoenergetic photon beams (Belgin, 2022). Recent study suggested the use of kilovoltage X-rays for determination of linear and mass attenuation coefficients (Chen et al., 2012). This can be achieved by calculating the effective energies of the kilovoltage x-rays by using the half-value layer (HVL) method.

The paraffin wax has similarity of physical properties to the synthesized bolus in radiotherapy treatments including the value of the relative electron density are equivalent to human soft tissue (Aisyah et al., 2020). The use of paraffin wax was considered to be an affordable, convenient, successful, and adaptable approach for reducing the air gap between human tissues during irregular superficial cancer irradiations (Pires et al., 2023).

This study measured the mass attenuation coefficients of paraffin wax and other commonly used solid phantoms by using the effective energies of kilovoltage X-rays. The measured mass attenuation coefficients of the materials were also compared to the simulated values by using XCOM software.

2. Methodology

2.1. Preparation of paraffin wax phantoms

The paraffin wax phantom was fabricated from solid crystalline pellets of paraffin wax as shown in Figure 1(a). The crystalline pellets were melted by double boiling method as its melting point is below the boiling point of water of 100ºC. The molten wax was then poured on a silicon square shaped mould with eternal dimensions of 18 cm x 18 cm x 1 cm. The molten wax in the mould was then cooled down and solidified at room temperature until a slab paraffin wax was achieved as in Figure 1(b).

Figure 1: (a) The crystalline pellets of paraffin wax and (b) the fabricated paraffin wax phantoms.

2.2. Determination of x-ray effective energies

The effective energy was determined using the half value layer (HVL) method for nominal x-ray energies of 60, 81 and 125 kVp based on the study by Chen et al., (2012). The focal spot-surface distance (FSD) was fixed at 100 cm and X-ray beam field size of 16 cm × 16 cm. A Shadow Free Diagnostic (SFD) ionisation chamber was placed on the table couch and connected to an electrometer to obtain the reading of the dose in Gy. The ionisation chamber was exposed and the initial reading of the dose was obtained. An aluminium (Al) sheet with thickness of 1 mm was then added under the collimator box of the X-ray unit. The Al sheets were subsequently added, with increment of 1 mm, until the electrometer readings reached half of the original value. The values between log X-rays transmission and Al thickness was plotted by using a semi-log graphs and the value of HVL of the Al were determined from the graph. The value of the linear attenuation coefficient was obtained from the graph while the mass attenuation coefficient of Al was calculated by calculating the ratio between linear attenuation coefficient and the density of Al. The effective energy of the X-rays was obtained by referring the XCOM software based on the value of the calculated mass attenuation coefficient (Chen et al., 2012).

2.3. Measurement of mass attenuation coefficient

The mass attenuation coefficient of paraffin wax, solid water phantoms, Perspex® and water were measured by using transmission method of X-rays as shown in Figure 2. The phantom materials were suspended between the x-ray tube and the ionisation chamber at nominal X-ray energies of 60, 81 and 125 kVp at automatic exposure control (AEC) set up. The transmitted X-rays from each exposure will be recorded and the thickness of phantoms and water were added at 2 cm increment on each exposure until a total of 10 cm thicknesses. When photon passing through a medium with thickness$x$, its intensity will be attenuated and reduced by the medium according to Beer-Lambert's law by the equation:

$$I = I_{\circ} e^{-\mu x} \qquad (1)$$

with $I_{\circ}$ is the initial intensity of photon, and μ is the linear attenuation coefficient of sample in cm^{-1}.The linear attenuation coefficient of sample can be calculated by modifying Equation (1) into

$$\mu = \frac{1}{x} \ln\left[\frac{I_{\circ}}{I}\right] \qquad (2)$$

The mass attenuation coefficient, μ/ρ (cm²/g) of sample can be obtained by dividing the value of linear attenuation coefficient with the density of the sample. The graphs of ln (I_o/ I) against thickness of samples were plotted for all samples at all X-ray energies. The value of linear attenuation coefficient, μ can be obtained by calculating the gradient of the graph at all photon energies based on the study by Marashdeh et al. (2012) and Yusof et al., (2017). The mass attenuation coefficients of phantoms and water were compared to the theoretical values by using XCOM at all experimental energies.

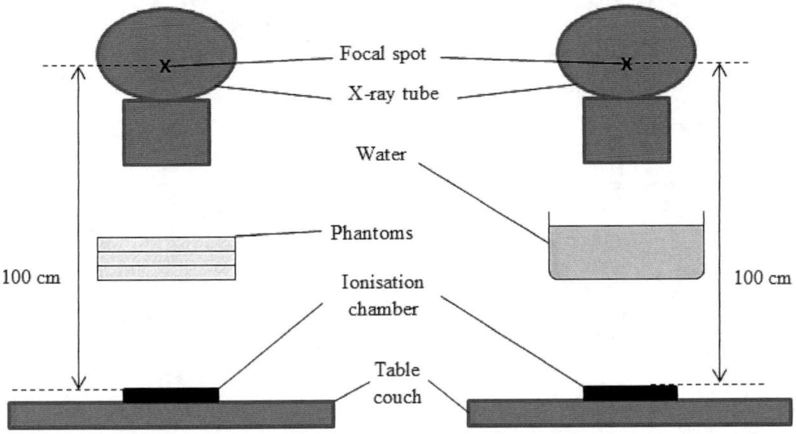

Figure 2. Experimental set up for the measurement mass attenuation coefficient by using X-ray effective energies in (a) phantoms and (b) water.

3. Results and discussion

The effective X-ray energies from the nominal X-rays energies are presented in Table 1. The calculated effective energies were in good agreement with the previous study by Chen at al., (2012). The graphs of log X-ray transmission, ln (I_o/ I) versus thickness of phantoms in all nominal X-ray energies were illustrated in Figure 3. From the figure, the transmission of X-rays in all phantoms material was linear with the increased thickness of phantoms. As the thickness of the material increases, the attenuation values decrease, indicating a reduced ability of the medium to absorb X-ray radiation (Budiawan et al., 2019). The linear regression value, R^2 that were close to the value of 1.0 indicated a good linearity between X-ray transmissions and the increased thickness of phantoms. From the equation (2) the gradient of the graph was recorded as the linear attenuation coefficient, μ of the phantom materials (Yusof et al., 2017).

The graph of mass attenuation coefficient of paraffin wax, Perspex® and solid water phantoms in comparison to water measured at all nominal effective X-ray energies is illustrated in Figure 4. The graph showed good agreement between solid water phantoms to water in all effective X-ray energies. The mass attenuation coefficients of paraffin wax and Perspex® however were found to be lower compared to water in lower X-ray effective energies and n good agreement to water and solid water at higher effective X-ray energies. The detail of mass attenuation coefficients of the phantoms and their percentage of differences to water at all nominal effective X-ray energies are listed in Table 3.

Table 1. The calculated effective X-ray energies by using HVL method

X-ray energy (kVp)	60	81	125
Effective Energy (keV)	35.22	37.57	41.20

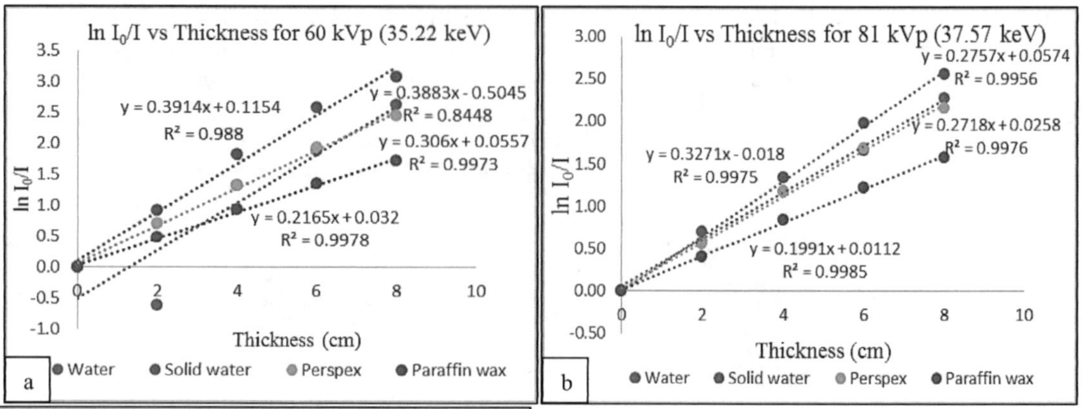

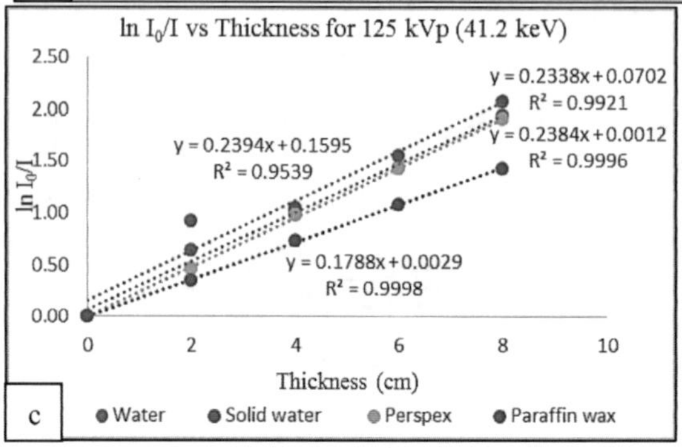

Figure 3. Graphs between X-ray transmission ln (Io/I) and thickness of phantoms for (a) 60 kVp, (b) 81kVp and (c) 125 kVp X-rays. The gradients of the graphs were used as linear attenuation coefficient of the phantoms

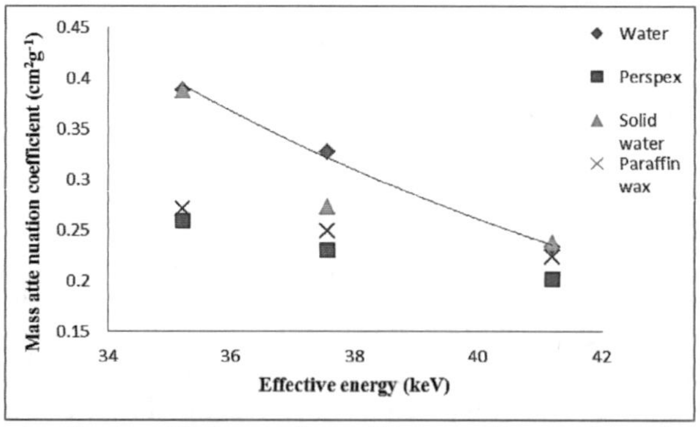

Figure 4. The mass attenuation coefficients of paraffin wax, Perspex® and solid water phantoms in comparison to water at all experimented effective X-ray energies.

The comparison between the measured mass attenuation coefficient and the XCOM values at similar photon energies for all experimented phantom materials are presented in Table 3. The results showed low percentage of differences in paraffin wax within 9% percentage of deviation followed by water within 26%. These uncertainties were noted due to the increased of backscattered photons from the phantom. According to Shukor et al. (2018), the amount of backscatter photons is due to the increased of field size used during the measurements. This is also consistent with the previous study by Nair et al., (2017) that suggested the uncertainties might be due to the wide beam geometry that covering the detector chamber, resulting to the detection of scattered photons as well. Nair et al., (2017) also reported that the distance between focus to chamber and distance from focus to attenuator also influenced the measurements accuracy. There are currently no standard approach for the comparison of mass attenuation of materials as tissue equivalent phantoms. A comparison to water is made as water is being proposed as standard phantom material in international dosimetric protocols.

Table 2. The detail values of mass attenuation coefficients of paraffin wax, Perspex® and solid water in comparison to water at all experimented effective X-ray energies.

Phantom	Density (g/cm^3)	Effective Energy (keV)	Mass attenuation coefficient (cm^2/g)	Percentage of Deviation with water (%)
Solid water	1.00	35.22	0.3871	0.3
		37.57	0.3271	19.8
		41.2	0.2370	1.3
Perspex	1.18	35.22	0.2593	33.2
		37.57	0.2303	15.6
		41.2	0.2020	13.6
Paraffin wax	0.80	35.22	0.2706	30.3
		37.57	0.2489	8.7
		41.2	0.2235	4.4
	1.0	35.22	0.3883	-
Water		37.57	0.3271	-
		41.2	0.2338	-

The results indicated the suitability of using x-ray effective energies as alternative method besides the monoenergetic photon energies for the measurement of mass attenuation coefficients. This method is more feasible by using available set up in the x-ray imaging facilities. The use of this method however is limited to the x-ray energy ranges available.

Table 3. The mass attenuation coefficients of several materials by using the x-ray effective energies in comparison to the XCOM calculations.

Phantom	Effective energy (keV)	Mass attenuation coefficient (cm^2/g)		
		Experimental	XCOM	Percentage of deviation (%)
Water	35.22	0.388	0.413	6.0
	37.57	0.327	0.373	12.3
	41.2	0.234	0.315	25.8
Solid water	35.22	0.387	0.557	30.5
	37.57	0.273	0.491	44.5
	41.2	0.237	0.404	41.4
Perspex	35.22	0.2593	0.336	22.9
	37.57	0.2303	0.315	27.0
	41.2	0.2020	0.276	26.9
Paraffin wax	35.22	0.271	0.280	3.6
	37.57	0.249	0.265	6.3
	41.2	0.223	0.244	8.4

4. Conclusion

The mass attenuation coefficients of paraffin wax, Perspex®, solid water phantom and water were able to be determined using effective energies of kilovoltage X-rays. The measured mass attenuation coefficients of the materials decreased as the X-ray effective energy increased. The comparison between the measured and XCOM values showed good agreement between the two methods at all experimented X-ray effective energies. The use of x-ray effective energies can be considered as alternative experimental method beside the monoenergetic photon energies. The overall results indicated the suitability of X-ray effective energies to be used for the measurement of mass attenuation coefficients of materials and to determine a new phantom materials for the use in medical physics works.

5. References

[1] Nair R B, Ramakrishnan G, Chandralingam S and Kurup P 2017 Evaluation of phantom equivalent materials in polychromatic diagnostic x-ray beam. *JMAHS*, **6(2)**; 24–30.

[2] Tekin H O, Sayyed M I, Erguzel T T, Karahan M, Kilicoglu O, Mesbahi A, and Kara U 2018 Investigation of water equivalence and shielding properties of different solid phantoms using MCNPX code. *Dig. J. Nanomater. Biostruct.*, **13(2)**; 551–562.

[3] Frimaio A, Nascimento B C, Barrio R M M, Campos L L and Costa P R 2019 X-ray spectrometry applied for determination of linear attenuation coefficient of tissue-equivalent materials. *Radiat. Phys. Chem.*, **160**; 89–95.

[4] Kargar S L, R., Abtahi S M M and Akbari M E 2019 Investigation of the radiological properties of various phantoms for their application in low energy x-rays dosimetry. *Radiat. Phys. Chem.*, **157**; 33–39.

[5] Marashdeh M W, Bauk, S, Tajuddin A A, Hashim R 2012 Measurement of mass attenuation coefficients of Rhizophora spp. binderless particleboards in the 16.59–25.26 keV photon energy range and their density profile using x-ray computed tomography. *Appl. Radiat. Isot.*, **70**; 656-662.

[6] Yusof M F M, Yahya M H, Rosnan M S, Abdullah R, and Kadir A B A 2017 Dose measurement using Al2O3 dosimeter in comparison to LiF:Mg, Ti dosimeter and ionization chamber at low and high energy x-ray. *AIP Conf. Proceed.*, **1799**; 040007.

[7] Mousa A, Kusminarto K and Suparta, G B 2017 A new simple method to measure the x-ray linear attenuation coefficients of materials using micro-digital radiography machine. *IJAER*, **12**; 10589–10594.

[8] Belgin E E 2022 Comparison of gamma spectrometric method and XCOM method in calculating mass attenuation coefficients of reinforced polymeric composite materials. *Radiat. Phys. Chem.*, **193**; 109960.

[9] Chen S C, Jong W L and Harun A Z 2012 Evaluation of x-ray beam quality based on measurements and estimations using SpekCalc and Ipem78 models. *Malays. J. Med. Sci.*, **19(3)**; 22–28.

[10] Aisyah S, Carina C C C, Nazara T, Sekartaji G, Nainggolan A and Endarko 2020 A comparative study of dosimetric characterization of bolus based on natural rubber (Hevea Brasiliensis) and clinical bolus for therapy with megavolt electron radiation. *J. Phys.: Conf. Ser.*, **1505** 012026.

[11] Pires A M, Borges F, Dias A G, Joana L, Matos M and Gonçalves J 2023 Paraffin gauze bolus as tissue compensator in photon irradiation for mycosis fungoides – regarding a case study. *JRP*, **22(e82)**; 1-4.

[12] Budiawan M A, Suryani S, Abdullah B and Tahir D 2019 Analysis of Absorption Properties of a Composite FlyAsh and Fe 2 O 3 for X-ray Radiation Shielding Applications. *IOP Conf. Ser.: Mater. Sci. Eng.*, **593(1)**; 012014.

[13] Shukor N S A, Asri N H M, Jaafar S M F S and Rosnan M S 2018 Measurement of backscatter factor for kilovoltage x-ray beam using ionization chamber and Gafchromic XR-QA2 film. *IOP Conf. Ser.: Mater. Sci. Eng.*, **298(1)**; 012043.

6. Acknowledgement

The authors would like to acknowledge the financial support from the USM Research Universiti (RU-I) grant scheme no 1001.PPSK.8012381. The authors also would like to acknowledge the Radiology Department, Hospital Universiti Sains Malaysia (HUSM) for the use of the X-ray machine and ionization chamber used in this study.

IOP Conf. Series: Materials Science and Engineering **1326** (2025) 012002 doi:10.1088/1757-899X/1326/1/012002

Preliminary study: Development of an automated data collection system for gamma ray measurements using Python

S.M. Yusuff[1*], M.A.N.M. Tamron[2], I. Mustapha[1], A.H. Ismail[1], M.F. Abdul Rahim[3], S.A. Idris[4], N.A. Abd Rahman[1], and N. Ramli[1]

[1]Malaysian Nuclear Agency, Bangi, 43000 Kajang, Selangor, Malaysia
[2]International Islamic University of Malaysia, 53100 Kuala Lumpur, Malaysia
[3]MAHSA University, Jalan SP 2, Bandar Saujana Putra, 42610 Jenjarom, Selangor, Malaysia
[4]University Malaysia of Computer Science and Engineering (UNIMY), VSQ@PJ City Centre, Jalan Utara, Section 14, 46200 Petaling Jaya, Selangor, Malaysia

*E-mail: soleha@nm.gov.my

Abstract. This paper presents an automated data collection system for gammaray measurement using the Python programming language. The system automates experimental data collection from gamma-ray detection systems, providing greater confidence in research results. The objective of this study is to develop a Python coding system for automated data collection and to produce significant graphical data results. The study used Python coding to continuously measure and analyze gamma rays emitted from Ba-133 radioactive material, which were detected using a scintillation detector and counted using a scaler ratemeter. The coding system used NumPy, Pandas, Matplotlib, and Tkinter Python packages, and Microsoft Visual Studio Community for integration. The Beer-Lambert attenuation law formula is embedded in the coding system to produce intensity against time graphical data results for various sample materials. The developed Python code was integrated into the gamma-ray detection system. The automated data collection system operates without significant lag or errors for 3 hours of gamma-ray measurements with 6 to 10 seconds of time interval speed. The graphical user interface (GUI) and data output are visualized immediately and continuously, and all data output can be safely stored on the computer. A Python-based automated data collection system was successfully developed, allowing efficient visualization of significant graphical data results. This system can reduce radiation exposure for radiation workers and may integrate with cloud computing for remote operation in the future.

1. Introduction

Gamma rays are crucial in medicine, industry, agriculture, scientific research, and security applications fields [1]. Their high energies (above 100 keV) and short wavelength range (less than 10 pm) enable them to rapidly interact with the electrons of materials [2]. Researchers used the gamma transmission method while developing the nuclear gauges. For instance, Beigzadeh *et al.* [2] engineered a nuclear densitometer using the gamma transmission method, featuring a thallium doped sodium iodide (NaI(Tl)) scintillation detector and Cs-137 radioactive source, to measure the density of 10 different types of woods ranging from 0.317 to 0.819 g/cm^3. Beigzadeh *et al.* [2] also proposed this nuclear densitometer for glasses, plastics, and other materials. When compared to conventional wood densitometers such as the volumetric method, pylodyn wood tester, torsiometer, extraction of nails driven into trees, and resistograph, the nuclear densitometer emerged as the most promising option due to its fast detection method, non-destructive method, and ability to provide precise, accurate, and reliable density results [2].

In contrast to Beigzadeh *et al.*, [2], Aung and Tun, [3] utilized the gamma transmission method, NaI(Tl) scintillation detector, and Cs-137 radioactive source to analyze the gamma radiation shielding properties (linear attenuation coefficient and mass attenuation coefficient) of five hardwoods and two softwoods from Myanmar's tropical wood species [3]. Beigzadeh *et al.*, [2] and Aung and Tun, [3] used the Beer-Lambert law attenuation equation (equations 1, 2, and 3) to determine the linear attenuation coefficient, mass attenuation coefficient, and density of various wood types, respectively.

$$I = I_0 e^{(-\mu x)} \tag{1}$$

$$I = I_0 e^{(-\mu_m \rho x)} \tag{2}$$

$$\rho = \frac{-1}{\mu_m x} ln \frac{I}{I_0} \tag{3}$$

where I is count of gamma ray detected by detector after pass through a sample, I_0 is count of gamma ray without a sample, μ is linear attenuation coefficient (1/cm), x is thickness of sample material (cm), μ_m is mass attenuation coefficient (cm^2/g), and ρ is density of sample material (g/cm^3).

In modern nuclear applications, the development of sophisticated instrumentation involves hardware, acquisition, processing, and software components [4]. Of these, software development for nuclear instruments is crucial for creating efficient and cost-effective software that meets user requirements [4]. Python is a versatile programming language widely utilized in data science, natural science, software development, Internet applications, and machine learning [5]. Python automates numerous essential operations performed at the processor level. Per Ryzhkov *et al.*, [5] Python's rapid development and widespread adoption can be ascribed to its open-source characteristics, straightforward syntax, diverse built-in data structures, a vast array of freely accessible libraries for nearly any purpose, and its capability to operate on multiple platform devices.

This paper aims to develop an automated data collection system for a simple gamma radiation-based instrument comprising a detector and a radiation source that follows Beer–Lambert's law to detect changes in one material. In a simple gamma-ray measurement system, the recorded signal detected by the Na(TI) scintillation detector is a known factor, while the changes of material amount (thickness) or material types (density) remain the unknown factor [6]. Thus, an automated data collection system was developed in this study using the Python programming language and Visual Studio Code as the development environment. Figure 1 shows the schematic diagram of the general setup for the simple gamma-ray measurement system using gamma transmission method.

IOP Conf. Series: Materials Science and Engineering **1326** (2025) 012002 doi:10.1088/1757-899X/1326/1/012002

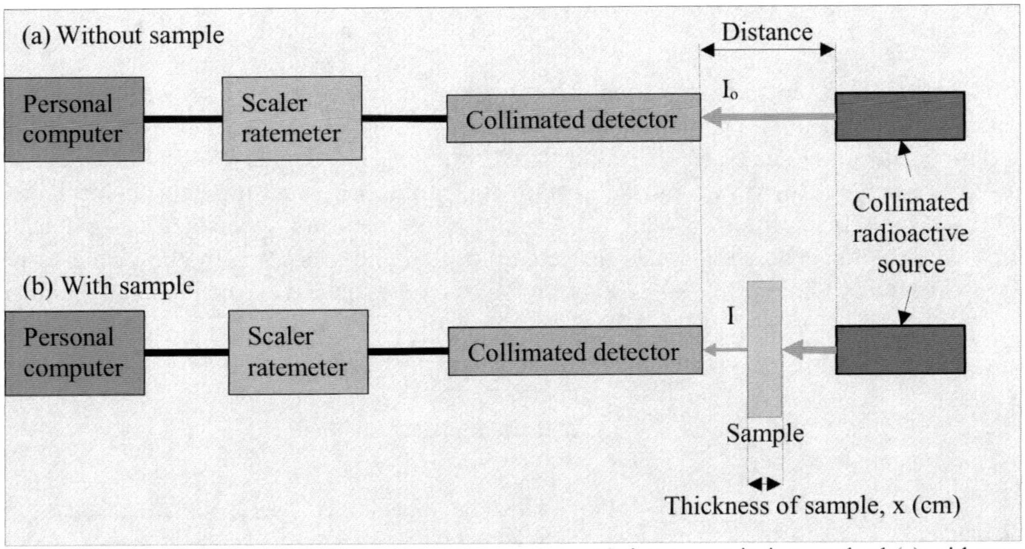

Figure 1. Experimental set-up of gamma ray measurement using transmission method (a) without and (b) with sample material.

2. Methodology

2.1. Hardware

A data collection system for measuring gamma rays was developed by utilizing a personal computer (PC) with a Windows operating system, Intel(R) Celeron (R) CPU 5205U@1.90GHz 20 GB installed RAM and a 64-bit operating system, a Ludlum Model 2200 scaler ratemeter (see Figure 2), a NaI(Tl) scintillator detector (Ludlum), an AC power supply cable, an RS-232 cable connector to USB, and a 5m detector cable connector to scaler ratemeter. The scaler ratemeter has a 9-pin RS-232 port for PC connection, front control settings for adjusting high-voltage, threshold, discriminator, and window, and is compatible with AC power supply and internal batteries. It can accommodate Geiger-Muller, proportional, and scintillator-type detectors.

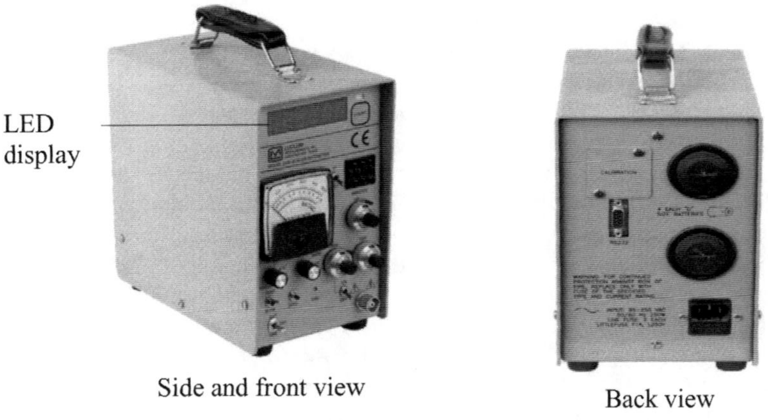

Figure 2. Scaler ratemeter used in this study.

IOP Conf. Series: Materials Science and Engineering **1326** (2025) 012002 doi:10.1088/1757-899X/1326/1/012002

This study utilized the Ludlum Model 2200 scaler ratemeter due to its versatility in measuring radioactive contamination and radiation. It is compatible with various detectors, including scintillation detectors (e.g., NaI(Tl)), proportional counters, and Geiger-Müller (GM) detectors. Figure 3 presents the hardware block diagram for gamma-ray measurement, comprising an NaI(Tl) detector, the Ludlum Model 2200 scaler ratemeter, and a personal computer. The gamma-ray detector generates signals that are amplified and processed by the single-channel analyzer within the Ludlum Model 2200, with the results displayed on the personal computer.

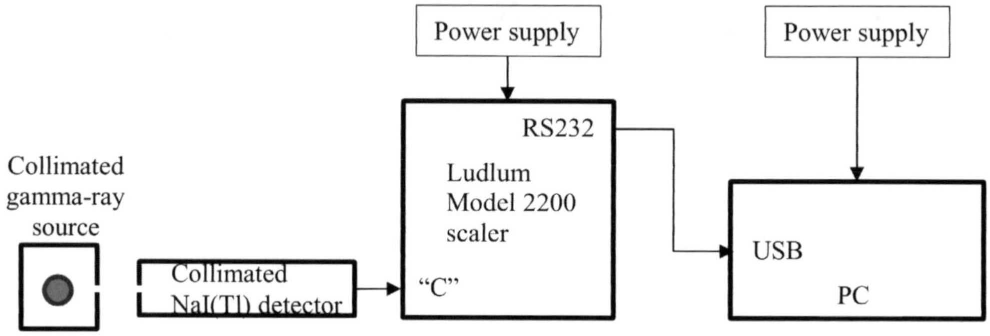

Figure 3. Hardware block diagram of the data collection system.

2.2. Programming code development

The data collection system has three main functions: 1) acquiring data, 2) calculating intensity, average, and density, and 3) generating graphs. The Visual Studio Code environment and the Python programming language, with the following libraries: Datetime, Serial, Threading, Numpy, Pandas, Tkinter, and Matplotlib, were used to develop the data collection system.

The DateTime library is essential for managing dates and times, while the Serial library facilitates communication between the PC and the scaler ratemeter. Additionally, the Threading library supports multi-threading within a single process, and the Numpy library handles large arrays, multi-dimensional arrays, and a range of mathematical functions for array operations. The Pandas library provides for data manipulation and analysis, the Tkinter library provides a standard GUI toolkit, and the Matplotlib library enables the creation of static, interactive, and animated visualizations, such as scatter plots.

Figure 4 depicts the process involved in the data collection system development and its distinct process architecture. This flowchart shows the development of a data collection system using Python programming in Visual Studio Code environment starting with initializing the GUI application using Tkinter followed by configuring the serial connection between the personal computer and scaler ratemeter of Ludlum Model 2200 scaler ratemeter through specific serial strings command ("SS" (start scanning) and "RS" (read scanning)). After that, code for set count time and read the count time were developed to ensure the scaler ratemeter count with a specific time using "SC" (set count) and "RC" (read count) strings command. Then, this study was developed the programming code to start communication between these two devices, followed by start thread, getting the background count, initial intensity, and intensity, opening graph window for three different graphs (Count vs Time, Intensity vs Time, and Density vs Time), updating data table, save data to the.CVS file, and stop the measurement/ stop communication with the scaler ratemeter. The developed data collection system for gamma-ray measurements using Python includes a comprehensive set of thirteen functions, represented by the following pseudocode.

IOP Conf. Series: Materials Science and Engineering **1326** (2025) 012002 doi:10.1088/1757-899X/1326/1/012002

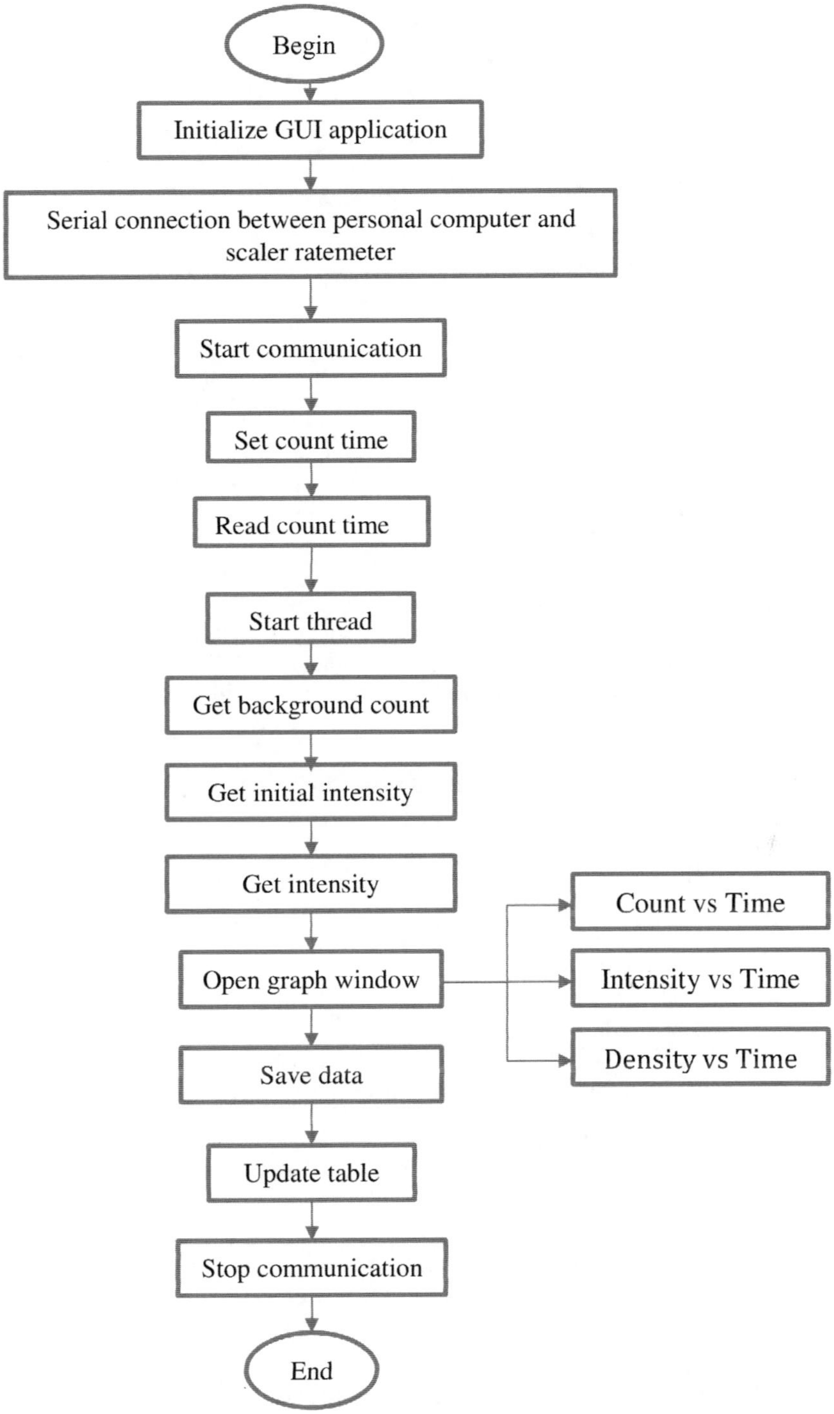

Figure 4. Flowchart of process in development of data collection system.

BEGIN

Initialize the GUI Application

CREATE a Tkinter application window

SET the title of the window to "Ludlum 2200 Scaler Ratemeter"

```
        CREATE a new frame widget
        CREATE a menubar and add the following:
                "File" menu with "Exit" option
                "Info" menu
        ARRANGE containers and frames for:
                Operator name
                Date
                Configuration
                Sample Information
                Average Count
                Attenuation Parameter
                Set Count Time
                Start/Stop/Save options
                Graph and Data Entry
        CREATE labels and entry fields for user input
        ADD functionality to open or close the "Information" window based on user actions
        GET the current date
        UPDATE the date label in the GUI
Serial Connection Functionality
        ENCODE the command string
        SEND the encoded command through the serial connection
        READ a line of data from the serial connection
        DECODE the received data into a string and remove whitespace
        RETURN the cleaned output
Start Communication
        PROMPT user for port, baud_rate, data_bits, parity, stop_bits, timeout_bit, and sleeptime
        INITIALIZE a serial connection
        SET is_sending to TRUE
        INITIALIZE an empty list data
        SEND "SS" command and RECEIVE the output
        IF the output is "OK":
                WAIT for sleeptime
                SEND "RS" command
                RECEIVE the next output
                APPEND the current time and output to data
                CALL the update_table function
        ELSE:
                DISPLAY an error message
Set Count Time
        GET count time from user input
        CREATE and SEND 'SC' command
        IF the response is "OK":
                DISPLAY a success message
        ELSE:
                DISPLAY an error message
        CLOSE the serial port
Read Count Time
        CREATE and SEND 'RC' command
        IF the response is received:
                DISPLAY "Count Time: [response] seconds"
        ELSE:
```

IOP Conf. Series: Materials Science and Engineering **1326** (2025) 012002 doi:10.1088/1757-899X/1326/1/012002

```
            DISPLAY an error message
Start Thread
        SET the thread to daemon mode and START it
        INITIALIZE variables for communication data, status, and the thread
Get Background Count
        SET the "Wait" text on the Background_Count_button
        SEND "SS\n" command
        IF the response is "OK":
            WAIT for sleeptime
            SEND "RS\n" command
            READ 10 counts
            CALCULATE the average background as the mean of the 10 counts
            DISPLAY the Background_Count
        ELSE:
            DISPLAY an error message
Get Initial Intensity
        SET the "Wait" text on the Initial_Count_button
        SEND "SS\n" command
        IF the response is "OK":
            WAIT for sleeptime
            SEND "RS\n" command
            READ 10 counts
            CALCULATE the average initial intensity as (mean of 10 counts - average back-
            ground)
            DISPLAY Io as the average initial intensity
        ELSE:
            DISPLAY an error message
Get Intensity
        SEND "SS\n" command
        IF the response is "OK":
            WAIT for sleeptime
            SEND "RS\n" command
            READ the count
            CALCULATE intensity as (count - average background)
            CALCULATE density as: −ln(intensity/average initial intesity)/mass attenua-
            tion coefficient×thickness)
            DISPLAY I as intensity and Density as density
        ELSE:
            DISPLAY an error message
Open Graph Windows
        FOR Count vs Time, Intensity vs Time, and Density vs Time:
            CREATE a new graph window
            SET the title (e.g., "Count vs Time")
            PREPARE DataFrame with appropriate columns (Time, Count, Intensity, or Den-
            sity)
            PLOT the graph with labels and title
            SCHEDULE periodic updates
Save File
        PROMPT user for a file path
        CREATE a DataFrame with columns ['Index', 'Time', 'Count', 'Density', 'I']
        SAVE the DataFrame in Excel format
```

```
    Stop Communication
            SET is_sending to FALSE
    Update Table
            CLEAR all rows from the table widget
            INSERT each row from data into the table
END
```

2.2.1. Configuration

The Ludlum Model 2200 scaler ratemeter interfaces with a personal computer via its RS-232 serial port for data logging. To ensure proper communication, the RS-232 parameters must be configured, and specific commands sent before requesting and receiving count data. The NaI(Tl) detector connects to the Ludlum Model 2200 scaler ratemeter through the detector input on the front panel, using a "C" coaxial connector. Additionally, the 9-pin "D" RS-232 connector on the back of the Ludlum Model 2200 scaler ratemeter is linked to the USB port of the personal computer using an RS-232-to-USB cable. This setup enables communication between the scaler ratemeter and the personal computer. At the RS-232 data dipswitch, the switch is set to "PC," and the count time push wheels are adjusted to "000." In "PC" mode, bi-directional data communication allows the personal computer to control the counting process and manage the count time of the Ludlum Model 2200 scaler ratemeter via the RS-232 port. The RS-232 parameters are configured within the GUI, setting the Baud Rate to 2400, Data Bits to 8, Parity to None, and Stop Bit to 1. With this configuration, the personal computer, equipped with the data collection system, can manage count time, start and stop counting operations, timestamp the data, and save the information using serial string commands.

2.2.2. Formula

The programming code includes the formulas for average background count (equation 4), average initial count (equation 5), initial intensity (I_o) (equation 6), intensity (I) formula (equation 7), and density formula (equation 2) [2,3]. It also takes input information for known I_0, μ_m, x, sample name, operator name, and count time. Additionally, the code generated graphs of count against time, intensity against time, and density against time.

$$Averagebackgroundcount = \frac{\sum_1^{10}(backgroundcount)}{10} \tag{4}$$

$$Averageinitialcount = \frac{\sum_1^{10}(initialcount)}{10} \tag{5}$$

$$Initialintensity(I_0) = Averageinitialcount - Averagebackgroundcount \tag{6}$$

$$Intensity(I) = Count - Averagebackgroundcount \tag{7}$$

2.2.3. Functions for GUI

The Tkinter library created the GUI and offered a visual representation of the essential user interface screens' layout and navigation. The GUI includes interface buttons for configuring RS-232 parameters, entering operator name and date, inputting sample information, setting attenuation parameters, displaying data arrays, enabling data logging, visualizing graphs, start and stop count, as well as saving the data in .CSV file.

The data collection system was tested using a Barium-133 gamma-ray emitter placed 10 cm in front of the NaI(Tl) detector, as illustrated in Figure 1(a). During the experiment, the high voltage (HV) dial setting was gradually adjusted in increments of 0.1 every 6 seconds while counting. This process continued until a sharp increase in results was observed following the plateau phase. The gamma-ray intensity was continuously recorded throughout the experiment.

IOP Conf. Series: Materials Science and Engineering **1326** (2025) 012002 doi:10.1088/1757-899X/1326/1/012002

3. Results and discussion

'Ludlum 2200 Scaler Ratemeter', the GUI created using Python in Visual Studio Code (Figure 5) in this study. It has a Menu bar, Setting panel, Data Entry, and Visual data parts. The GUI acts as a human-machine interface, enabling users to interact seamlessly with the Ludlum Model 2200 scaler ratemeter and a personal computer using a mouse for intuitive control. The main window, dialog boxes, frames, menus, icons, buttons, checkboxes, and other graphical elements in the designed GUI are utilized to display information and provide user-friendly controls.

In this study, the development of a GUI for material identification based on varying densities is presented. Users can enter sample thickness and attenuation parameters into the GUI and obtain the gamma-ray count data from the scaler ratemeter as recorded by the GUI. It uses coded formulas (Equations 1–7) for analysis and visualization. Users can view data through plots of count, intensity, and density versus time for materials with different thicknesses and attenuation coefficients.

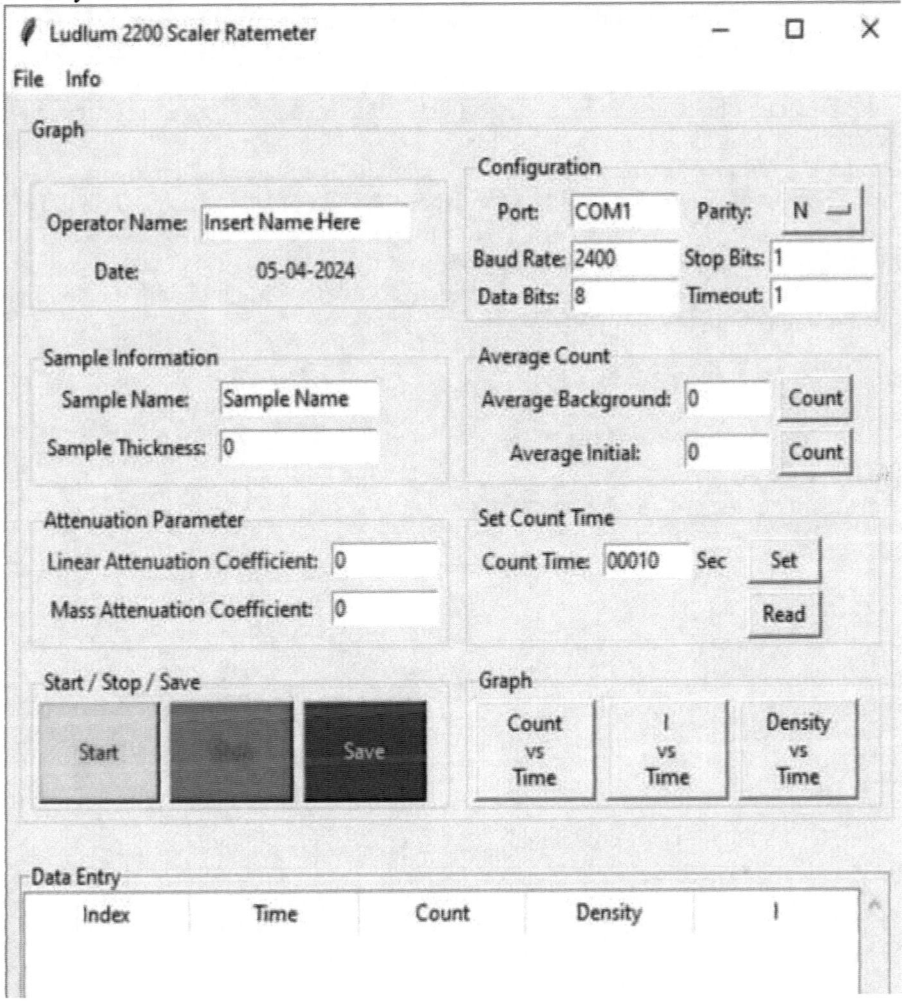

Figure 5. The developed GUI of 'Ludlum 2200 Scaler Ratemeter'.

In practical, after setting up the hardware, users able to open the GUI of 'Ludlum 2200 Scaler Ratemeter' in the personal computer and configure it by inputting the Baud Rate to 2400, Data Bits to 8, Parity to None, and Stop Bit to 1. The interface also allows users to create a new project by entering the operator name, sample name, sample thickness, and set count time. The date and time displayed on the interface are similar to the computer's system time. The interface enables users to start gamma ray measurements, stop gamma ray measurements, and save data in table form in the .CSV file. Additionally, users can input name, sampel name, thickness of sample, count time, and attenuation coefficient informations. Users can use button 'Count' at Average count to obtain average background count and average initial intensity. From the GUI, users also able to view real-time count data and observe graphs of Intensity vs. time, Count vs. time, and Density vs. time. The start_communication, set_count_time, read_count_time, get_average_background, get_average_initial, open_graph_window, open_graph_window1, and open_graph_window2 functions were coded into smaller blocks for multiple times execution without repetitive coding.

The interface workflow of an automated data collection system for simple gamma-ray measurement has interface design parts and external functions parts (graphs). The user interface design and the graphs could be synergized into one file because the workflow uses one programming language [8]. Splitting the interface structure into two parts improves the readability and amendability of codes. The major Python packages are Tkinter for interface design, Pandas for data manipulation, and matplotlib for visualizing processed results. The developed data collection system meets the user's specifications and operates continuously for 3 hours without errors.

Figure 6 illustrates the output of the GUI, displaying a graph of the measured gamma-ray intensity plotted against time. The data was collected as the HV dial was incrementally adjusted from 0.0 to 3.5, corresponding to approximately 0 V to 500 V of high voltage (HV). The graph demonstrates an increase in the gamma-ray intensity detected by the detector, followed by a plateau. This plateau is a critical result for stability test as a quality control procedure for detectors before use to obtain the appropriate operating voltage and stable output reading [9].

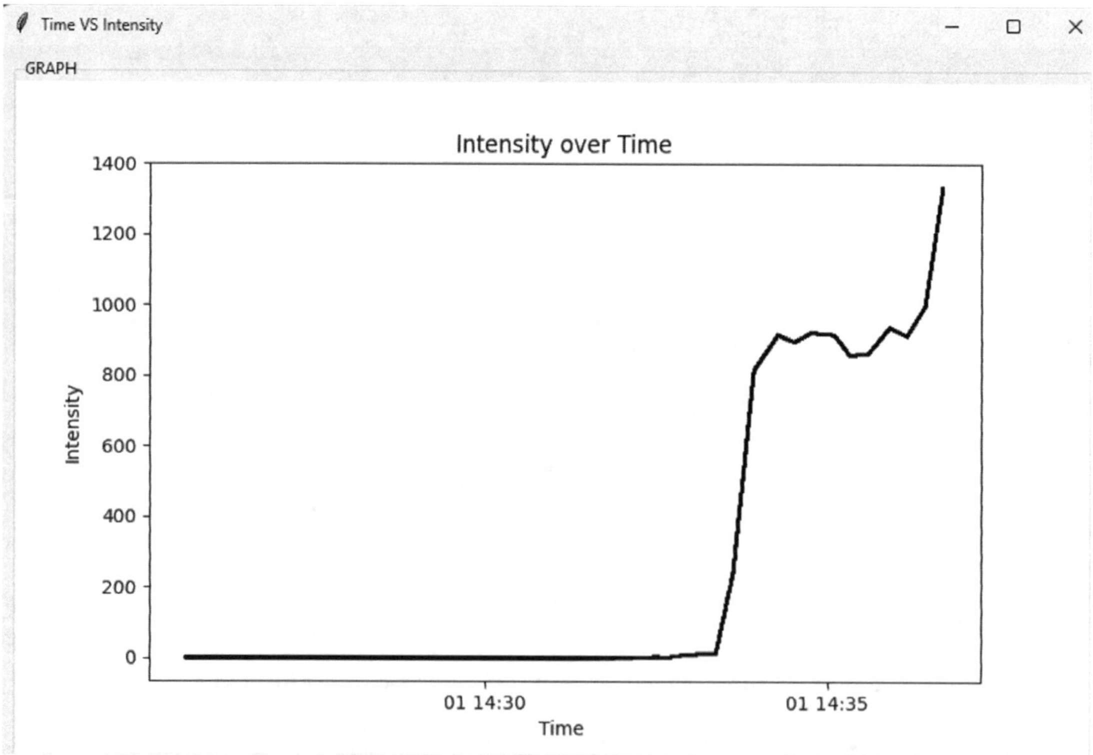

Figure 6. The data visualization of measured gamma-ray intensity against time with HV increasing.

4. Conclusion

The data collection system for gamma-ray measurement was successfully developed in this study using Python programming language and Visual Studio Code. This open-source software offers cost-effective development of a data collection system. In conclusion, the gamma-ray data collection system significantly enhanced safety when handling radioactive sources during research, development, and technical services activities. This study strongly recommends integrating the data collection system with cloud computing to enable mobile support and remote system operation.

5. References

[1] AlzahraniJ S, Alalawi A, Alrowaili ZA, Sriwunkum C and Al-Buriahi 2024 Gamma attenuation and buildup factors of GeO_2-Ga_2O_3–Gd_2O_3 transparent glass ceramics for shielding applications *J.of Radiat. Res.and Appl. Sci.* **17**(4) 101073.

[2] Beigzadeh A M, Vaziri M R R, Soltani Z and Afarideh H 2019 Design and improvement of a simple and easy-to-use gamma-ray densitometer for application in wood industry *Measurement* **138** 157.

[3] Aung Y Y and Tun KT 2019 Investigation of gamma radiation shielding characteristics for various wood slabs by using gamma attenuation method *Int. Res. J. of Adv. Eng. and Sci.* **4**(3) 99.

[4] Gaytan-Gallardo and Rivero-Gutierrez 2007 IAEA-TECDOC-1565-Validation procedures of software applied in nuclear instruments *Proc. of a Technical Meeting* (Vienna).

[5] Ryzhkov F V, Ryzhkova Y E and Elinson M N 2023 Python in chemistry: physicochemical tools *Processes* **11**(10) 2897.

[6] Alamoudi M, Sattari M A, Balubaid M, Eftekhari-Zadeh E, Nazemi E, Taylan O and Kalmoun E M 2021 Application of gamma attenuation technique and artificial intelligence to detect scale thickness in pipelines in which two-phase flows with different flow regimes and void fractions exist. *Symmetry* **13**(7) 1198.

[7] Palani N 2020 One-GUI designing for medical devices & IoT introduction. *Trends Dev. Med. Devices* 17.

[8] Chen J, Zou Y, and Chu X 2024 DIFFUSUP: A graphical user interface (GUI) software for diffusion modeling. *Appl. Computing and Geosci.* **22** 100157.

[9] Susiapan Y, Rahim R A, Zain R M, Wahab A R, Wahid H, Ishak M H I, Bunyamin S, Yunus Y, and Ling L P 2015 Stability test for NaI(Tl) scintillation detector. *Jurnal Teknologi (Sciences & Engineering)* **77**(17) 27.

Acknowledgement

The authors are thankful to Muhammad Asyraf Noorza Tamron, who participated in the Practical Industry program at the Malaysian Nuclear Agency under this PreQualification of Research and Development (PQRD) project (NM-R&D-23-10), Malaysian Nuclear Agency, and all supporting staff (Mohamad Rabaie Shari, Airwan Affandi Mahmood, Hearie Hassan, Ainul Mardhiah Terry@Aldeline Terry Ngau, and Nurliyana Abdullah) from the Plant Assessment Technology Group of the Industrial Technology Division of the Malaysian Nuclear Agency. In addition, authors would also like to acknowledge Tabung Akaun Amanah of the Malaysian Nuclear Agency for giving some financial support.

IOP Conf. Series: Materials Science and Engineering **1326** (2025) 012003 doi:10.1088/1757-899X/1326/1/012003

Self-attenuation correction in environmental bulk samples: a study on enhancing gamma-ray activity measurement using HPGe detector with soil samples in varied geometries

Sahar Saleh Al-Horr[1], Khoo Kok Siong[1,2,*]

[1]Nuclear Science Programme, Department of Applied Physics, Faculty of Science and Technology,Universiti Kebangsaan Malaysia, 43600 Bangi, Selangor, Malaysia
[2]Nuclear Technology Research Centre (NUKLEAR), Faculty of Science and Technology, Universiti Kebangsaan Malaysia, 43600 Bangi, Selangor, Malaysia

*Corresponding Author: khoo@ukm.edu.my

Abstract. Gamma-ray spectrometry is a vital technique for identifying and measuring radioactivity in environmental samples, offering high energy resolution. However, the accuracy of measurements in bulk samples is often compromised due to geometric and compositional variations from reference samples, leading to self-attenuation effects. This study aims to improve the accuracy of gamma-ray activity measurements in bulk soil samples by applying a self-attenuation correction across various geometric shapes and sample matrices. The correction methodology was implemented by comparing the photon attenuation within the sample to a standard air sample, taking into account the linear attenuation coefficient and sample thickness. This method was experimentally verified using High Purity Germanium (HPGe) detectors, resulting in improved accuracy of activity concentration measurements by 3.88% to 6.72%. The proposed method provides a practical alternative to existing techniques, including LabSOCS simulations and empirical corrections by offering a practical, adaptable solution without the need for extensive computational resources. The self-attenuation correction reduced measurement discrepancies by 4.3% to 10.8%, particularly in samples with significant density variations. Beyond environmental monitoring, this correction methodology has potential applications in other fields requiring precise radioactivity measurements, such as medical physics, industrial radiography, and nuclear safety protocols. These findings underscore the critical importance of accurate self-attenuation corrections for reliable environmental and radiological monitoring.

1. Introduction

Gamma-ray spectroscopy is one of the most commonly used techniques for detecting the concentration of natural and synthetic radionuclides in environmental samples [1]. The use of high-purity germanium (HPGe) detectors in high-resolution gamma-ray spectroscopy enables precise quantitative determination of radioactive concentrations in various sample types. This is due to the advantages of HPGe detectors, such as high sensitivity, accurate radiation detection, the ability to measure radioactivity in samples of various sizes and shapes, and the low radioactive background during measurements.

IOP Conf. Series: Materials Science and Engineering **1326** (2025) 012003 doi:10.1088/1757-899X/1326/1/012003

However, one of the critical challenges in gamma-ray spectrometry is accurately determining the number of photons absorbed by the detector [2]. The response of the HPGe detector is strongly influenced by the size, shape, and density of the sample being analyzed [3]. Bulk samples are often preferred for their potential to yield more representative measurements, despite their associated challenges [4], their complex geometric shapes can introduce significant attenuation effects, reducing the number of photons reaching the detector and leading to inaccurate measurements [5].

Accurate determination of radionuclide concentrations using gamma-ray spectroscopy is often hindered by self-attenuation effects, particularly in bulk samples of varying shapes and sizes. These effects reduce the number of photons reaching the detector, leading to inaccuracies in radiometric measurements [2, 5]. Existing techniques for self-attenuation correction, such as empirical methods or computational simulations like LabSOCS [6], may either oversimplify the problem or require complex simulations that are difficult to implement in routine applications [7]. Therefore, there is a need for a practical and accurate method to correct for self-attenuation in order to enhance the precision of gamma-ray spectrometric analysis, especially for environmental monitoring and other applications involving bulk samples [8].

To address this challenge, self-attenuation corrections were applied in this study by calculating the attenuation of gamma-ray photons through the sample material, considering the samples linear attenuation coefficient (μ) and thickness (x), following the relation:

$$I = I_0 \, e^{-\mu x} \tag{1}$$

where I_o denotes the intensity (peak count rates) without attenuation, I the intensity (peak count rates) after attenuation, μ the linear attenuation coefficient (cm^{-1}) and x denotes the sample thickness (cm).

To compute the attenuation of a certain substance compared to a standard air sample, the intensity ratio between the air sample and the substance sample is provided by the following equation:

$$e^{-\mu x} = \frac{I}{I_0} \tag{2}$$

The following equation may be used to compute self-attenuation correction factors for cylindrical samples with height x and linear attenuation coefficient (μ) (where μ= 0) for a standard air sample:

$$F_{att} = \frac{\mu x}{1-e^{-\mu x}} \tag{3}$$

By taking the logarithm of equation (2) and then substituting the value of m in equation (3) the self-attenuation correction factor for the sample, F_{att} is expressed by the following equation:

$$F_{att} = -\frac{\ln(\frac{I}{I_0})}{1-(\frac{I}{I_0})} \tag{4}$$

Where I and I_0 are the peak count rates for the actual sample and the standard air sample, respectively. In this experiment, the self-attenuation correction was applied by comparing the count rates for each sample with those of a standard air sample. This method allowed us to account for the material-specific attenuation properties and improve the accuracy of gamma-ray spectrometric measurements.

IOP Conf. Series: Materials Science and Engineering **1326** (2025) 012003 doi:10.1088/1757-899X/1326/1/012003

While traditional methods for self-attenuation correction often rely on empirical formulas or simplified assumptions, our approach incorporates a more detailed analysis of the samples geometric properties and the linear attenuation coefficient. Existing methods, such as LabSOCS simulations [6], are widely used but may require extensive computational resources and assume ideal conditions. In contrast, our technique emphasizes practical correction factors that can be applied more readily in field-based and laboratory measurements, reducing the reliance on simulations. Furthermore, unlike some empirical methods [7]. our approach provides better adaptability for samples of varying shapes and sizes.

Though this study focuses on environmental samples, the self-attenuation correction method described here has broader potential applications. For instance, it can be applied to radiometric analysis in medical physics, where accurate dosimetry is critical, or in industrial applications, such as material testing and quality control, where precise measurements of radioactive materials are essential.

2. Experimental work

2.1. *Gamma-ray spectrometry*

In this work, a p-type HPGe detector (CANBERRA model GC3018) equipped with a CANBERRA model 2002CSL preamplifier, having a relative efficiency of 30%, and a multichannel analyzer with 16382 channels were used for measurements. The crystal has a diameter of 59 mm and a length of 50 mm. The detector energy resolution (FWHM) at 1.33 MeV for ^{60}Co is 1.8 keV. The detector is connected to a computer where gamma spectrum analysis is performed using the Genie-2000 gamma-ray spectroscopy software by Canberra.

2.2. *Samples preparation*

Fifteen soil samples were collected from five different locations around the UKM Nuclear Science Building. In the lab, the soil samples were dried at 80 °C for 72 hours. The dried soil was then ground using a specialized electric grinder and passed through a 0.5 mm mesh sieve. The samples were divided into three groups, each with a different geometric shape (with heights of 4.4, 6.5, and 10 cm, and diameters of 7, 5.5, and 6.5 cm, it was given codes Cy4.4, Cy6.5, Cy10 respectively) for each location. Each soil sample was sealed to achieve secular equilibrium. The standard IAEA Soil-375 was also prepared in three geometric shapes similar to the samples, with different weights, then labelled, sealed, and stored.

2.3. *Methodology*

Four gamma-ray point sources (^{241}Am, ^{133}Ba, ^{137}Cs, ^{60}Co) with the following energy transitions (59.54, 80.99, 302, 356, 661, 1173, and 1332 keV), were used to perform transmission measurements for air, soil, and standard samples of different geometric shapes. The soil samples were placed above the detector, and then the four-point sources were positioned directly above the samples. With a counting time of 1000 s, and the resulting spectra were collected and corrected for dead time. This procedure was repeated for both the standard and air samples. The linear attenuation coefficients (μ) and the self-attenuation correction factors (F_{att}) for these samples were determined using equation (2) and equation (4).

The activity concentrations of ^{226}Ra, ^{232}Th, and ^{40}K were determined by counting each soil sample and the standard IAEA Soil-375 samples in its three geometric shapes for 12 hours using an HPGe detector. The system was calibrated with a multi-nuclide standard. The photopeaks used in this analysis were 351 keV for ^{214}Pb and 609 keV for ^{214}Bi to determine the concentration of ^{226}Ra. For ^{232}Th, the photopeaks were 911 keV for ^{228}Ac, 583 keV for ^{208}Tl, and 238 keV for ^{212}Pb. The photopeak used for ^{40}K was 1460 keV.

IOP Conf. Series: Materials Science and Engineering **1326** (2025) 012003 doi:10.1088/1757-899X/1326/1/012003

The activity concentrations of naturally occurring radioactive material (NORM) were calculated using equation (5).

$$C = \frac{M_{std} \times A_S}{M_s \times A_{std}} C_{std} \qquad (5)$$

Where C and C_{std} denote the concentration of radionuclide (Bq/kg), M_s and M_{std}, refer to the mass (kg), while A_s and A_{std} refer to the net count of both sample and standard respectively.

3. Results and discussions

Table 1 displays the linear attenuation coefficient (μ) corresponding to each energy for three geometries of containers. The observed decrease in the linear attenuation coefficients with increasing energy is consistent with the basic principles of gamma ray interactions, especially the photoelectric effect, Compton scattering, and pair production that prevail in different energy ranges. The rise in linear attenuation coefficients with bigger containers can be attributed to the greater sample thickness through which the gamma rays must transit, increasing the likelihood of interaction figure 1. The greatest uncertainties are observed at the upper and lower ends of the energy range; for example, at 59.54 keV in the container with the longest length of 10 cm, the standard deviation is 0.036, indicating greater variability in the measurements; in contrast, at intermediate energies of 302 keV in the container with the length of 6.5 cm, the standard deviation is 0.019, indicating more consistent measurements.

Table 1. Average value of the linear attenuation coefficient (μ) for three geometric shapes of bulk soil samples for different photon energies

	Linear attenuation coefficient μ (cm^{-1})		
Energy (keV)	[a]Cy4.4	[b]Cy6.5	[c]Cy10
59.5	0.224 ± 0.018	0.265 ± 0.028	0.319 ± 0.036
80.1	0.165 ± 0.034	0.208 ± 0.030	0.237 ± 0.029
302	0.105 ± 0.034	0.136 ± 0.019	0.151 ± 0.019
356	0.080 ± 0.036	0.132 ± 0.018	0.144 ± 0.018
661	0.045 ± 0.013	0.127 ± 0.016	0.133 ± 0.016
1173	0.031 ± 0.010	0.110 ± 0.010	0.128 ± 0.013
1332	0.017 ± 0.016	0.112 ± 0.014	0.122 ± 0.013

[a]Cy4.4: Geometric shape of the sample with a length of 4.4 cm and a diameter of 7 cm.
[b]Cy6.5: Geometric shape of the sample with a length of 6.5 cm and a diameter of 5.5 cm.
[c]Cy10: Geometric shape of the sample with a length of 10 cm and a diameter of 6.5 cm.

These differences may be due to statistical variation in photon counts, changes in detector efficiency at different energies, or geometric influences of the sample. The R^2 value of 0.68 suggests that energy and container size have a substantial role in gamma-ray attenuation.

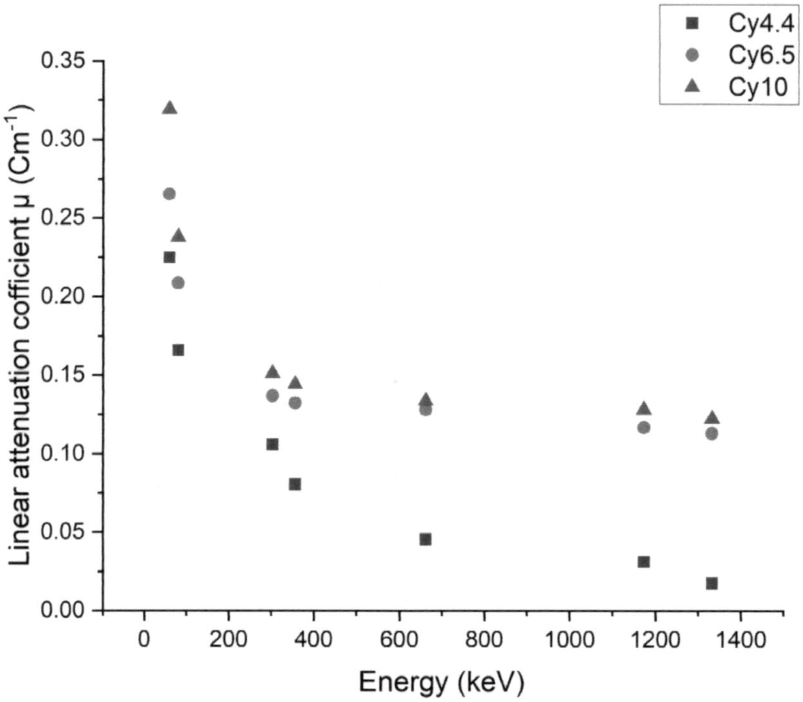

Figure 1. Variation of the Linear Attenuation Coefficient (μ) with Photon Energy for Different Sample Geometries

Self-attenuation correction is critical for detecting radioactivity with HPGe detectors because attenuation from the material between the source and the detector reduces measurement accuracy. Self-attenuation correction tries to increase the accuracy of radioactivity measurements by accounting for effects produced by photon absorption within the sample itself. Table 2 shows that the self-attenuation correction factor (F_{att}) decreases with increasing energy for three samples with varying geometric lengths. F_{att} values begin high at 59.54 keV energy and steadily drop as energy increases, reaching 1332 keV. This is because the self-attenuation effect is stronger at low energies due to increased photon absorption in the material, which diminishes as energy increases.

Furthermore, it is observed that the self-attenuation factor rises with increasing sample weight. At 59.54 keV energy, the F_{att} for a sample weighing 285±10 g (Cy10) is 2.638±0.226, compared to 1.241±0.021 for a sample weighing 83.58±5.93 g (Cy4.4) as in figure 2. This suggests that larger containers with more material have a stronger influence on photon absorption, resulting in a higher self-attenuation coefficient. The difference in self-attenuation, on the other hand, can be related to variations in sample density and chemical composition [7]. However, it can adjust the self-attenuation factor of samples with varied compositions, ignoring differences in chemical composition and density between reference and other bulk samples [6].

These findings highlight the need of considering self-attenuation when analyzing radiation data, particularly when utilizing HPGe detectors for precise measurements in samples with varying geometries and weights.

Table 2. Self-attenuation correction factors (F_{att}) for bulk soil samples with different weights

	Cy4.4		Cy6.5		Cy10	
Energy (KeV)	Sample weight (g)	F_{att}	Sample weight(g)	F_{att}	Sample weight(g)	F_{att}
59.5		1.241±0.021		1.659±0.083		2.638±0.226
80.1		1.175±0.038		1.502±0.083		2.146±0.170
302		1.123±0.022		1.314±0.048		1.672±0.010
356	83.58±5.93	1.083±0.039	106±3	1.303±0.046	285 ±10	1.637±0.100
661		1.046±0.014		1.292±0.040		1.584±0.082
1173		1.031±0.010		1.265±0.034		1.556±0.065
1332		1.017±0.016		1.255±0.035		1.529±0.063

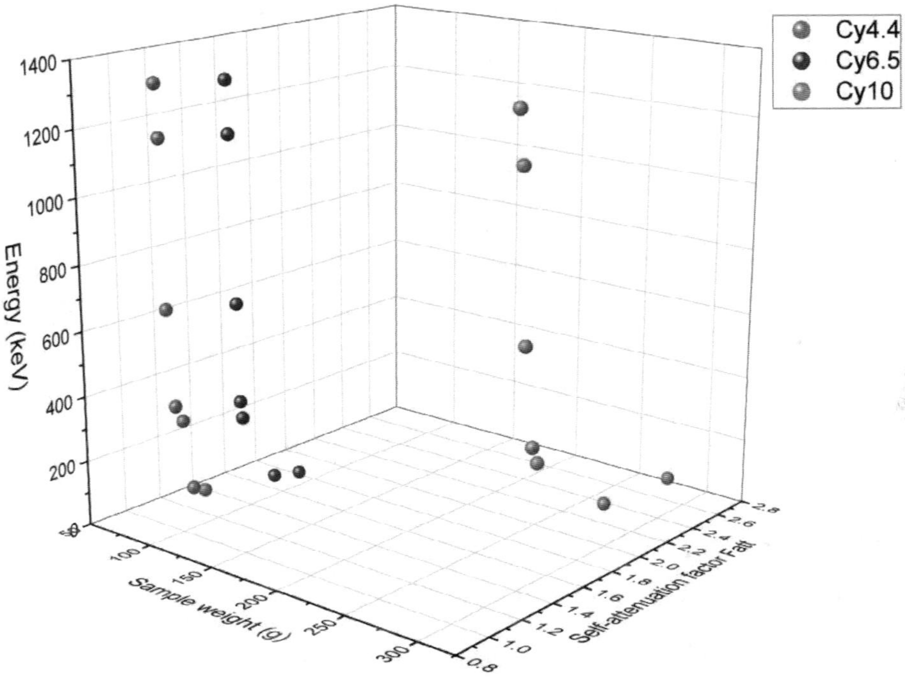

Figure 2. Variation of self-attenuation correction factors (F_{att}) with sample weight and absorbed photon energy

The activity concentration for (NORM) in soil samples with varied geometric shapes (lengths of 4.4, 6.5, and 10 cm) was determined using equation (5) and the HPGe detector, table 3. The results reveal that the measured radioactivity concentrations vary greatly depending on the geometric form of the sample. The biggest sample (Cy10) had the greatest concentration of ^{226}Ra detected at 30.71±0.55 Bq/kg, while the lowest concentration in sample (Cy6.5) was 26.08±3.00Bq/kg. This variance can be ascribed to differences in sample size and density, which impact photon absorption inside the material, as well as a lack of adjustment for self-attenuation.

IOP Conf. Series: Materials Science and Engineering **1326** (2025) 012003 doi:10.1088/1757-899X/1326/1/012003

The activity concentrations of ^{232}Th in the three samples were relatively close, ranging from 49.88±5.02 Bq/kg to 53.40±4.30 Bq/kg, indicating that the effect of self-attenuation on this element may be less clear than on other elements, or that the different geometric shapes of the samples have no significant effect on the accuracy of ^{232}Th measurement. Compared to other nuclides, ^{40}K shows a greater variation in activity concentration, with values ranging from 226±6 Bq/kg to 379±2 Bq/kg. The geometric shape of the sample influences the activity concentration of ^{40}K, as does the fact that the non-homogeneous distribution and lower total density of larger samples reduce the effect of self-attenuation, allowing more photons to reach the detector.

Table3. Average activity concentrations of ^{226}Ra, ^{232}Th, and ^{40}K for three geometric shapes of bulk soil samples

	Cy4.4	Cy6.5	Cy10
Radionuclides	Activity concentration (Bq/kg)		
^{226}Ra	28.26±4.18	26.08±3.00	30.71±0.55
^{232}Th	53.40±4.30	50.97±4.86	49.88±5.02
^{40}K	266±6	365±3	379±2

This highlights the importance of correcting for self-attenuation effects to obtain accurate estimates of activity concentrations in environmental samples [9]. Consequently, this study proposes an equation to adjust the calculated radioactivity values

$$C = \frac{M_{std} \times A_S \times F_{att,s}}{M_s \times A_{std} \times F_{att,std}} C_{std} \tag{6}$$

Where $F_{att,std}$ and $F_{att,s}$ the self-attenuation correction factors for the standard and the sample under study, respectively.

This equation provides precise estimates of radionuclide activity concentrations in soil samples with different geometric forms and densities by accounting for differences in sample and standard mass, net count rates, and self-attenuation correction factors. By taking into account these variables, it guarantees accurate measurement by resolving disparities brought on by variations in composition and attenuation. In order to improve measurement reliability across various sample geometries and efficiently minimize errors related to self-attenuation, the equation was used in this study to adjust the activity concentrations of radionuclides.

Equation (6) can be applied to environmental monitoring, where precision in radiation measurements is crucial. As a result, adding this correction to the observed activity via equation (6) will improve the assessment of environmental risks and the determination of radioactive contamination levels.

Table 4 presents the average activity concentrations corrected for NORM. The results indicate a notable decrease in the concentrations of ^{226}Ra ,^{232}Th and ^{40}K following correction, figure 3. This variation is attributed to the self-attenuation effect, where some gamma photons are either absorbed or scattered within the sample before reaching the detector. The initial measurements, which did not account for self-attenuation, led to inaccuracies in concentration estimates. The correction provides a more accurate representation of the number of photons emitted, resulting in improved measurement accuracy. Specifically, the accuracy improvements are 6.72%, 6.55%, and 5.34 % for ^{226}Ra, 6.66%, 6.33%, and 5.23% for ^{232}Th, and 5.74%, 3.88%, and 4.2% for ^{40}K for Cy 4.4, Cy 6.5, and Cy 10 containers, respectively. These results underscore the significance of accounting for self-attenuation in radioactivity measurements.

Self-attenuation transpires when gamma-ray photons are absorbed or dispersed within the sample, resulting in an overestimation of activity in uncorrected measurements. After applying the correction, the overestimations are decreased, resulting in lower activity concentrations across all geometries.

Self-attenuation is less important in smaller samples, such as Cy4.4 and Cy6.5, since photons have a shorter journey through the material and hence lose fewer particles. However, the correction accounts for the limited attenuation, resulting in a noticeable reduction in activity concentrations. In larger samples, the probability of gamma-ray photons essentially "colliding" with atoms through interactions like Compton scattering or pair production increases exponentially. This process, known as Compton scattering and dominant at intermediate photon energies results in energy loss, direction change of the incoming photons increasing a probability for higher-order absorption or subsequent scatter occurring with the sample. At higher photon energies, pair production becomes more important leading to the formation of electron-positron pairs which remove escaping photons from the initial radiation flow. Consequently, larger samples require more significant corrections to account for these interactions and improve measurement accuracy. While the correction accounts for Cy10 high photon losses, the decrease is less evident in comparison to smaller samples since the bigger volume increases the overall number of detected photons.

Table 4. Average corrected activity concentrations of ^{226}Ra, ^{232}Th, and ^{40}K for three geometric shapes of bulk soil samples

	Cy4.4	Cy 6.5	Cy10
Radionuclides	Corrected activity concentration (Bq/kg)		
^{226}Ra	26.36±4.16	24.37±2.57	29.07±0.40
^{232}Th	49.84±4.18	47.74±4.40	47.27±5.04
^{40}K	251±5	351±3	363±1

Figure 3 depicts the activity concentrations of ^{226}Ra, ^{232}Th, and ^{40}K before and after applying the self-attenuation correction. Where applying this correction reduced the measurement discrepancies by 7.53%, 10.8% and 4.3% for ^{226}Ra, ^{232}Th and ^{40}K respectively for the three geometries. These reduced underscores the significance of considering self-attenuation effects, especially for radionuclides such as ^{232}Th, where the most substantial discrepancy reduction was noted.

IOP Conf. Series: Materials Science and Engineering **1326** (2025) 012003 doi:10.1088/1757-899X/1326/1/012003

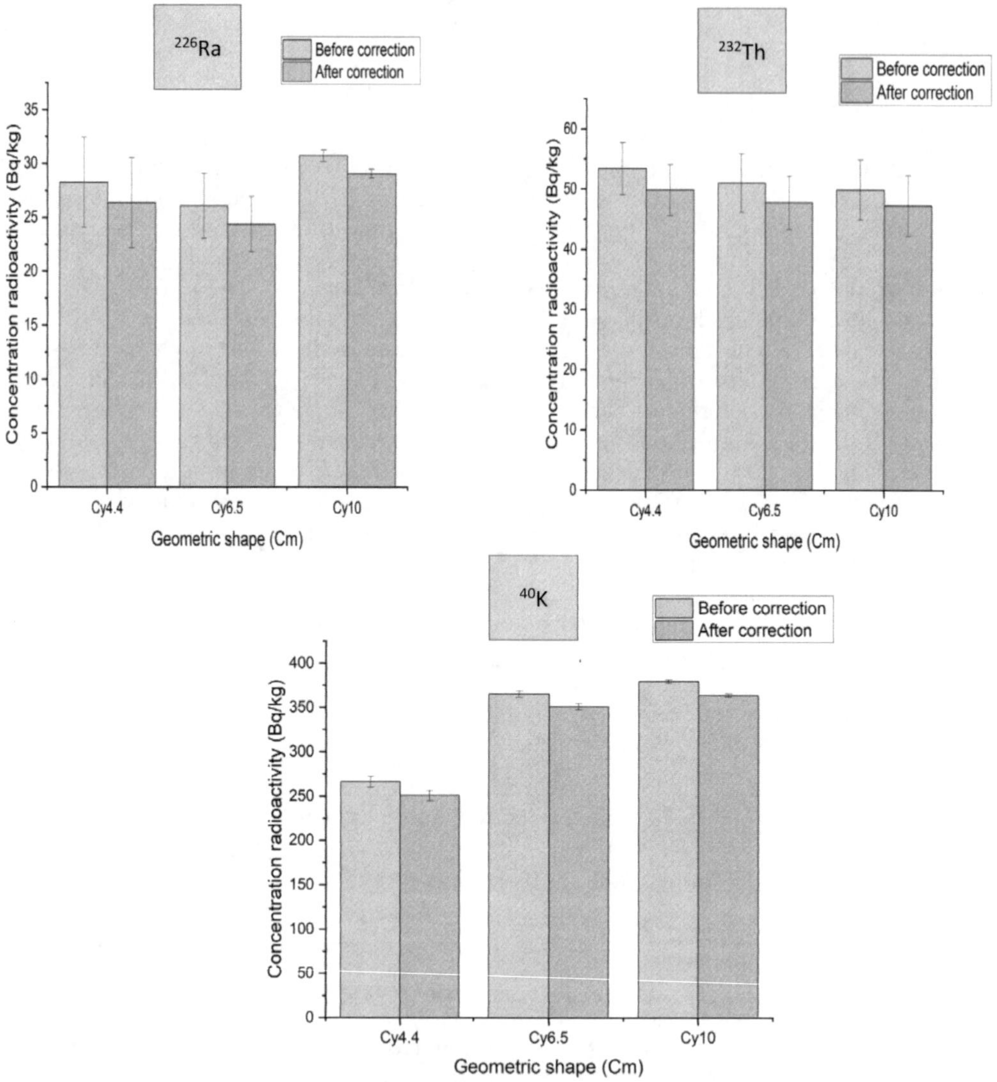

Figure 3. Activity concentrations of ^{226}Ra, ^{232}Th and ^{40}K before and after attenuation correction for three geometric shapes of samples.

The t-test was performed to compare variations in the improved of activity concentration accuracy after applying self-attenuation correction. The p-value for ^{226}Ra,^{232}Th ^{40}K was less than 0.05, indicating that the activity concentration accuracy for these radionuclides was improved, resulting in a significant difference between the data before and after correction. While the correlation coefficients show a clear relationship between geometric shape and radioactivity, the regression analysis's lack of statistical significance (p-value>0.05) demonstrates that geometric shape is not a direct or significant factor influencing radioactivity at the significance level used.

4. Conclusions

This study developed a self-attenuation correction method to improve gamma-ray spectrometry accuracy in bulk environmental samples. The method enhanced radioactivity measurements for 226Ra, 232Th, and 40K, showing improvements of up to 6.72%.

Future work should test this method on other sample types, such as water and industrial waste, and explore variations in material composition and photon energy. While focused on specific soil geometries, this approach is practical and effective for environmental monitoring and nuclear safety applications.

Compared to existing methods, this solution is accurate, efficient, and resource-friendly, supporting better radioactive contamination assessments and advancing gamma-ray spectroscopy techniques.

Acknowledgments

The authors would like to express their gratitude to Universiti Kebangsaan Malaysia (UKM) for providing the facilities and resources necessary for this research. Special thanks are extended to the Nuclear Technology Research Centre (NUKLEAR) and the technical team for their invaluable support and guidance. This research was partially supported by GUP-2023-019, with specific aspects of the study conducted on a small scale within the scope of this funding.

References

[1] Kaintura S S, Thakur S, Devi T 2024 Characterization and performance of co-axial HPGe detectors *J. Radioanal. Nucl. Chem.* **333** 3123–35 https://doi.org/10.1007/s10967-024-09376-y.

[2] Hafızoğlu N 2024 Efficiency and energy resolution of gamma spectrometry system with HPGe detector depending on variable source-to-detector distances *European Phys. J. Plus* **139** (2).

[3] Napoli E, Stenberg V Y, Juzeniene A, Hjellum G E, Bruland S, Larsen R H 2020 Calibration of sodium iodide detectors and reentrant ionization chambers for ^{212}Pb activity in different geometries by HPGe activity determined samples *Appl. Radiat. and Isot.* **166**.

[4] Andrade C 2020 Sample size and its importance in research *Indian J. of Psychol. Med.* **42**(1) 102–3.

[5] Kang M Y, Sun G M, Kim J, Choi H D 2016 Determination of HPGe peak efficiency for voluminous gamma-ray sources by using an effective solid angle method *Appl. Radiat. and Isot.* **116** 69–76.

[6] Tian Z N, Ouyang X P, Liu Y, Chen L, Liu J L, Zhang X P, Song J W, Zeng M 2014 Self-attenuation corrections calculated by LabSOCS Simulations for gamma-spectrometric measurements with HPGe detectors *Chinese Phys. C* **38**(7).

[7] Khater A E M, Ebaid Y Y 2008 A simplified gamma-ray self-attenuation correction in bulk samples *Appl. Radiat. and Isot.* **66**(3) 407–13.

[8] Badawi M S, Gouda M M, Nafee S S, El-Khatib A M, El-Mallah E A 2012 New analytical approach to calibrate the co-axial HPGe detectors including correction for source matrix self-attenuation *Appl. Radiat. and Isot.* **70**(12) 2661–8.

[9] Robu E, Giovani C 2009 Gamma-ray self-attenuation corrections in environmental samples *Romanian Reports in Phys.* **61** 295-300.

Compatibility evaluation and FTIR characterization of solvent-based paint enhanced with aerogel

A H Ismail[1*], S M Yusuff[1], N H Yusoff[1], M A Amir[2], S A Muhamad[3]

[1]Plant Asessment Technology Group, Industrial Technology Division, Malaysian Nuclear Agency, 43000, Kajang Selangor
[2]Biopolymer Group, Radiation Processing Technology Division, Malaysian Nuclear Agency, 43000, Kajang Selangor
[3]Radiopharmaceutical Technology Group, Medical Technology Division, Malaysian Nuclear Agency, 43000, Kajang, Selangor.

*hambali@nm.gov.my

Abstract. This study investigates the effects of incorporating aerogel into solvent-based paint formulations, focusing on appearance and Fourier Transform Infrared Spectroscopy (FTIR) characterization. Aerogels are known as high porosity materials, making them ideal for enhancing material thermal properties. In this study, solvent-based paint was incorporated with different concentrations of aerogel (1%, 3%, 5% and 7%). The mixtures were coated on the plate and evaluations were made on their appearance to ascertain the effect of adding aerogel to the solvent-based paint. It can be stated that the viscosity of the paint is directly proportional to the amount of aerogel added to it. FTIR analysis was conducted to study the molecular and structural changes due to the presence of aerogel. FTIR profile revealed that paint containing 1%, 3% and 5% silica aerogel produces a similar profile to the commercial paint formulation and paint containing 7% or silica aerogel produces new peaks at 1071 cm^{-1} indicating the presence of Si-O-Si bond to conclude the incorporation of silica aerogel into solvent-based paint.
Keywords. Aerogel; FTIR; paint; appearance

1. Introduction

Industries in Malaysia consist of many sectors such as petrochemical, fertilizer, chemical process, pharmaceutical and many more. This type of industry mainly consists of high-temperature process areas such as distillation columns, boilers, high-temperature furnaces, process vessels and many more. Whenever conducting non-destructive testing (NDT) in this industry, the high-temperature area will always be the obstacle faced by the testing personnel. This situation may affect the testing equipment such as the gamma radiation detector. It is not recommended that the detector be operated in high-temperature conditions. This is because the temperature can affect the count reading [1]. Another concern is on the functionality of the electronic parts of the equipment which can be reduced and then will reduce the shelf life of the equipment. Hence, applying a layer of insulation may be a positive step in reducing the side effects caused by the high temperature of the equipment. However, a layer of coating may not provide significant results as a heat resistant, therefore, some addition of material is needed to improve heat resistivity. Several materials are identified and produced positive results as heat insulation materials such as fibreglass, rock wools, cellulose, polymeric materials such as polystyrene and polyurethane, aluminium foil and ceramic coatings [2]. Another promising material that produced positive results as an insulation material is silica aerogel.

Aerogel is a promising material and does offer positive outcomes as a material to boost heat-resistant capability. Aerogel is a gel that consists of a microporous solid and proved to be the material that existed with low densities (0.003 – 0.15 kg/m3) with low density nano-porous solid with fine, open pore structure, larger surface area (500 – 1000 m2/g) [3]. Other than that, aerogel is also a material that has low thermal conductivity and excellent transparency [4]. Aerogel exists with low thermal conductivity characteristics due to its high porosity and nanometre pore size [5]. These characteristics of aerogel provide a wide range of aerogel applications in thermal and acoustic applications in building application sectors [6].

Furthermore, silica aerogel is also widely used in several sectors as an absorbent, nuclear waste storage, shock absorber, batteries and also as a catalyst [6]. Therefore, the potential to add aerogel into the coating is worth exploring. Previous study shows that the thermal conductivity of the acrylic emulsion coating improved after being reinforced with silica aerogel [7]. Other than that, a previous study also shows that the addition of aerogel to the water-based paint improved thermal conductivity and also reduced temperature transmission [8][9]. However, a limitation of water-based paint is its coating application surface. Solvent-based paint provides broader applications compared to water-based paint. By considering the industrial situation such as surface materials and conditions, solvent-based paint is more reliable. Hence, in this study, the compatibility of the mixture the aerogel and solvent-based paint will be assessed.

2. Material and Methodology

This study was conducted and divided into two main parts. The first part refers to the formulation of the solvent paint/aerogel mixture. The second part is the characterization part which represents the effects of adding aerogel to the solvent-based paint. The process flow in this study is depicted in Figure 1.

2.1. Material (aerogel, solvent paint)

Silica aerogel with brand ENOVA® Aerogel Particles with product code IC3120 was purchased from supplier Cabot. The aerogel is in clear white powder form with a particle size range of 0.1 to 1.2 mm size and with a pore diameter of approximately about 20 nm. The solvent-based paint, 9000 Gloss Finish was purchased and used the Nippon Paint brand from the existing commercial market in Malaysia.

2.2. Preparation of paint/aerogel mixture and appearance observation

The solvent-based paint was mixed with Enova silica aerogel in the normal room condition. 300 grams of solvent paint was poured into a container. The solvent paint was stirred vigorously using an electronic mixer while incorporating the Enova silica aerogel incrementally to obtain a uniform mixture. This step is critical to ensure the Enova silica aerogels were well dispersed with the solvent paint. The amount of Enova silica aerogel added to the solvent-based paint was (1%, 3%, 5%, and 7%) by weight. The appearance of the mixture was observed and the finding was recorded.

2.3. Characterization (Energy Dispersive X-Ray Fluorescent (ED-XRF), Fourier Transform Infra-Red (FTIR))

The element composition and purity of the purchased Enova silica aerogel were verified by using the ED-XRF (Shimadzu: ED-X-7000). The study on the chemical structure and the functional group was conducted by using FTIR (Agilent: Cary 630 FTIR Diamond ATR). All sample mixtures of solvent-based paint and Enova aerogel (1%, 3%, 5% and 7%) were assessed using FTIR including the original formulation of solvent-based paint and the purchased powdered Enova silica aerogel.

IOP Conf. Series: Materials Science and Engineering **1326** (2025) 012004 doi:10.1088/1757-899X/1326/1/012004

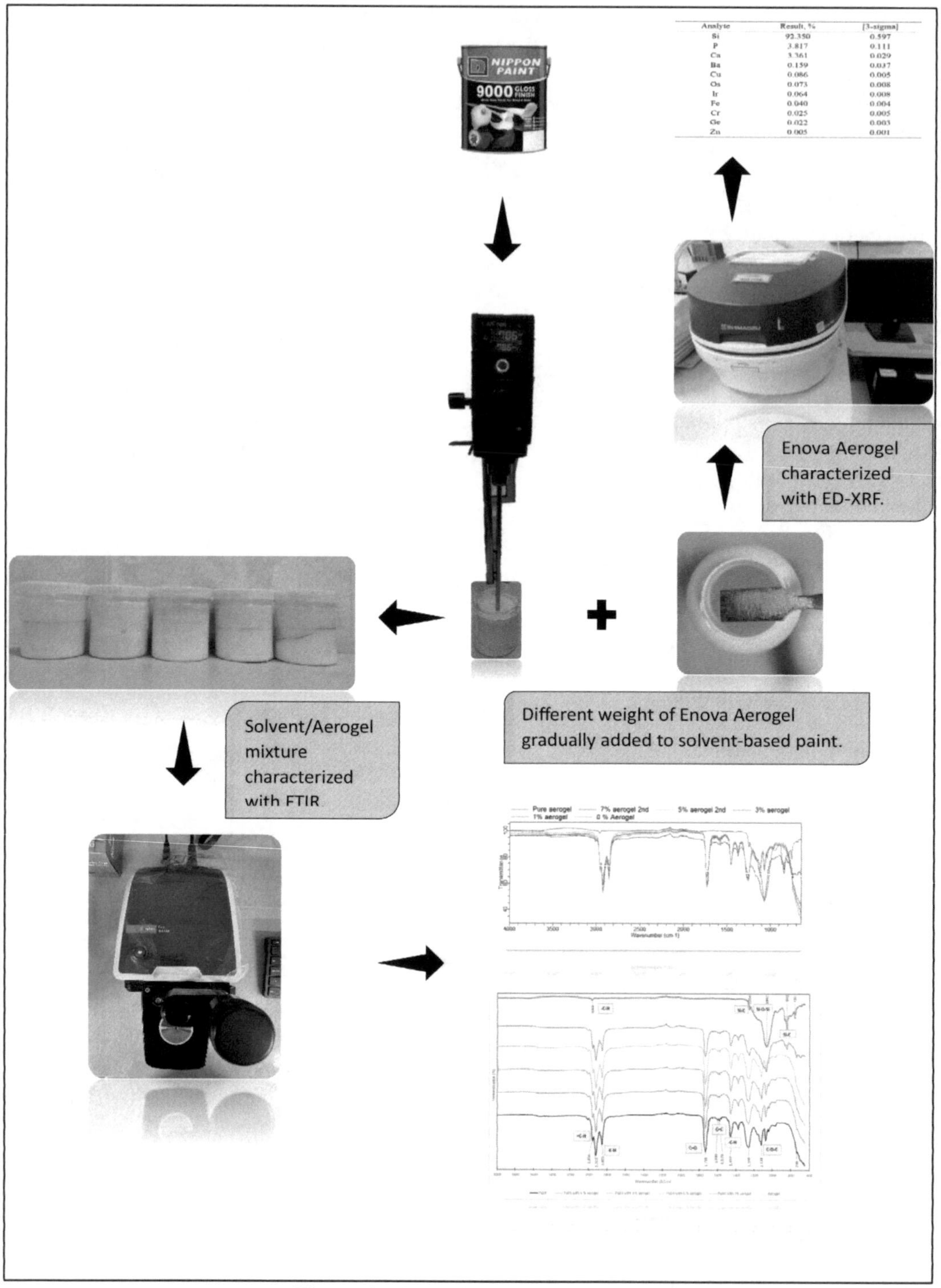

Figure 1. Flow of methods in this study.

3. Result and Discussion

The formulation of solvent-based paint and Enova aerogel mixture had been done at room temperature. Figure 1 shows the flow process conducted in this study. The purity of the Enova silica aerogel had been verified using the ED-XRF. Table 1 shows the purity test result of the aerogel. The Enova silica aerogel used in this study is composed of a Silica composition above 92% purity.

Table 1. Composition of the Enova Aerogel detected by ED-XRF.

Analyte	Result, %	[3-sigma]
Si	92.350	0.597
P	3.817	0.111
Ca	3.361	0.029
Ba	0.159	0.037
Cu	0.086	0.005
Os	0.073	0.008
Ir	0.064	0.008
Fe	0.040	0.004
Cr	0.025	0.005
Ge	0.022	0.003
Zn	0.005	0.001

The preparation process of solvent-based paint and aerogel mixtures was conducted with vigorous stirring conditions. To prevent any undispersed and undissolved particles, the Enova Aerogel was added gradually into the container depending on the ratio of the aerogels used. By using the original formulation of solvent-based paint as standard, it shows that the addition of the Enova silica aerogel into the solvent-based paint was driving the viscosity of the solvent paint to increase and the mixture became thicker. Figure 2 shows the effect on the flow properties due to the addition of Enova silica aerogel into the solvent-based paint. The newly mixed paints were also painted on the metal plate and the appearance was evaluated. The smooth surface of the paint became rougher due to coarse particles of Enova silica aerogel. It can be stated that the roughness of the surface appearance is directly proportional to the ratio of aerogel added. As the ratio of the Enova silica aerogel increased, the surface coat became rougher. This condition was similar to the viscosity of the solvent-based paint. This condition was considered due to the Enova silica aerogel remaining in solid particles within the solvents.

IOP Conf. Series: Materials Science and Engineering **1326** (2025) 012004 doi:10.1088/1757-899X/1326/1/012004

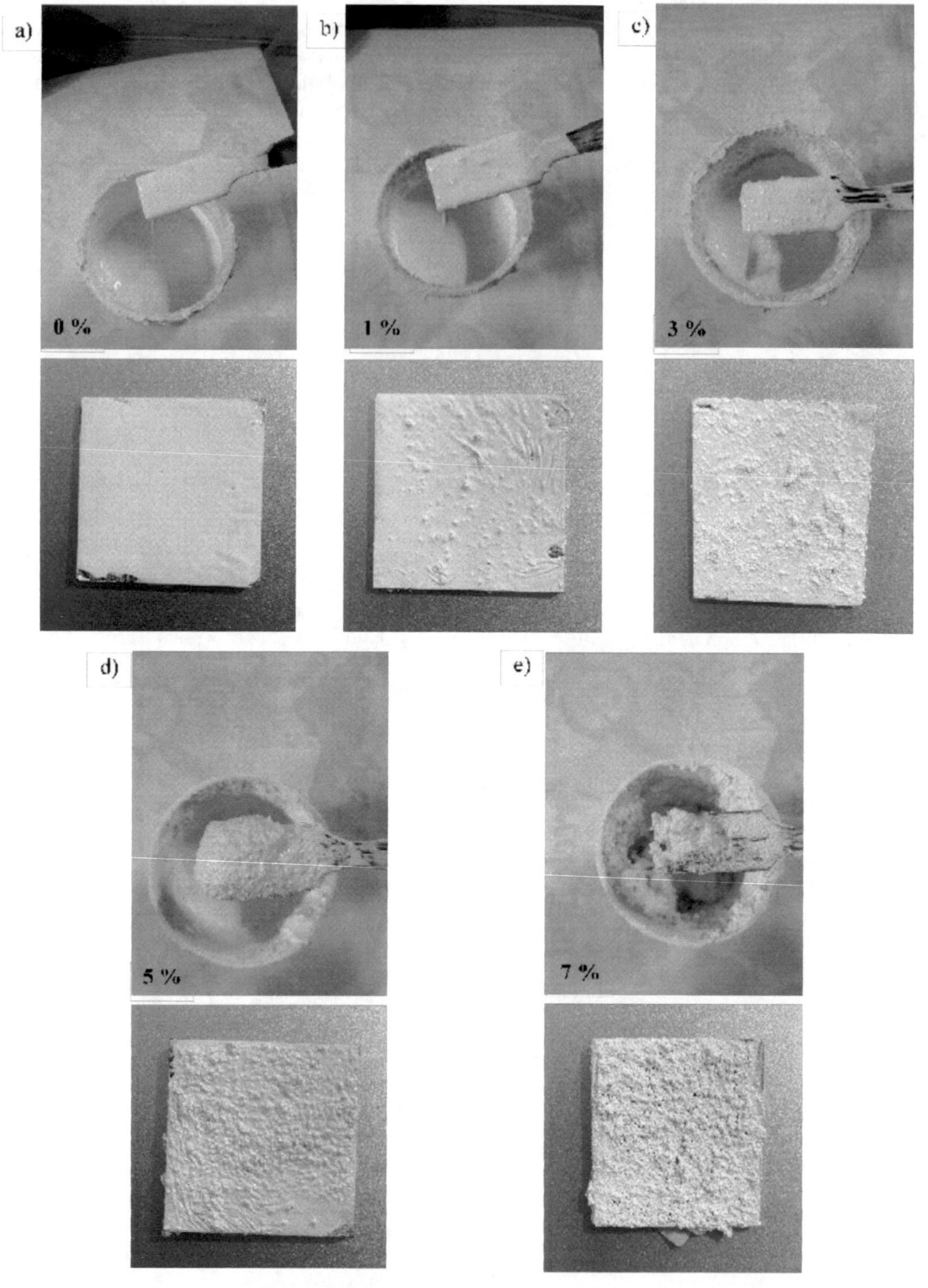

Figure 2. Observation after mix with (a) 0%, b) 1%, c) 3%, d) 5%, e) 7%) of Enova silica aerogel with the solvent-based paint and coating surface appearance on metal plate.

The infrared spectra (Figure 3) illustrate the functional group's presence in the commercial solvent-based paint, paint containing Enova silica aerogel (1%, 3%, 5%, and 7%), and Enova silica aerogel. The commercial solvent-based paint shows moderate absorption peaks at 2922 and 2853 cm^{-1}, attributed to the -C-H vibrations in the compound [10]. Additionally, absorption peaks at 1735 and 1120 cm^{-1} indicate the presence of C=O stretching vibrations in the carbonyl group [10][11] and C-O-C stretching of the ester group in the paint [11]. The band observed at 1457 cm^{-1} corresponds to -C-H bending vibrations [11]. The band at 2956 cm^{-1} referring to the unsaturation =C-H and the aromatic C=C stretching band at 1599 - 1579 cm^{-1} proved the alkyd compound in the commercial solvent-based paint [11][12][13].

The Enova silica aerogel exhibits absorption peaks at 844 and 1254 cm^{-1}, corresponding to Si-C, 1071 cm^{-1} indicating the presence of the Si-O-Si functional group, and 2960 cm^{-1} indicating the presence of C-H. These results are consistence with the FTIR spectrum of silica aerogel prepared by Ha-Yoon Nah [14] and superhydrophobic silica aerogels prepared using MTMS/TMOS with a molar ratio: of 75/25 precursor mixture [15].

FTIR profile of paint containing 1%, 3%, and 5 % silica aerogel similar to the commercial solvent-based paint. This finding revealed that the Enova silica aerogel from 1% to 5% content did not affect the functional groups of the commercial solvent-based paint. These results show that 1, 3, and 5 % of Enova silica aerogel content was too little to be detected by FTIR, and no chemical interaction between functional groups of the commercial solvent-based paint and Enova silica aerogel with 1, 3, and 5 % content.

In the case of paint containing 7% Enova silica aerogel, the absorption peaks ranging from 1165 to 4000 cm^{-1} closely resemble those of commercial solvent-based paint. However, the new peaks observed at 844 and 1256 cm^{-1} attributed to the Si-C bond, 1071 cm^{-1} corresponded to the Si-O-Si bond, and below 757 cm^{-1} appeared in the FTIR spectrum of paint containing 7% Enova silica aerogel. This particular bond is responsible for the presence of silica aerogel. Moreover, the peak at 1120 cm^{-1} corresponds to C-O-C stretching of the ester group has disappeared. These peak changes are rigid evidence that 7% silica aerogel is sufficient to modify paint with chemical interaction between paint and Enova silica aerogel.

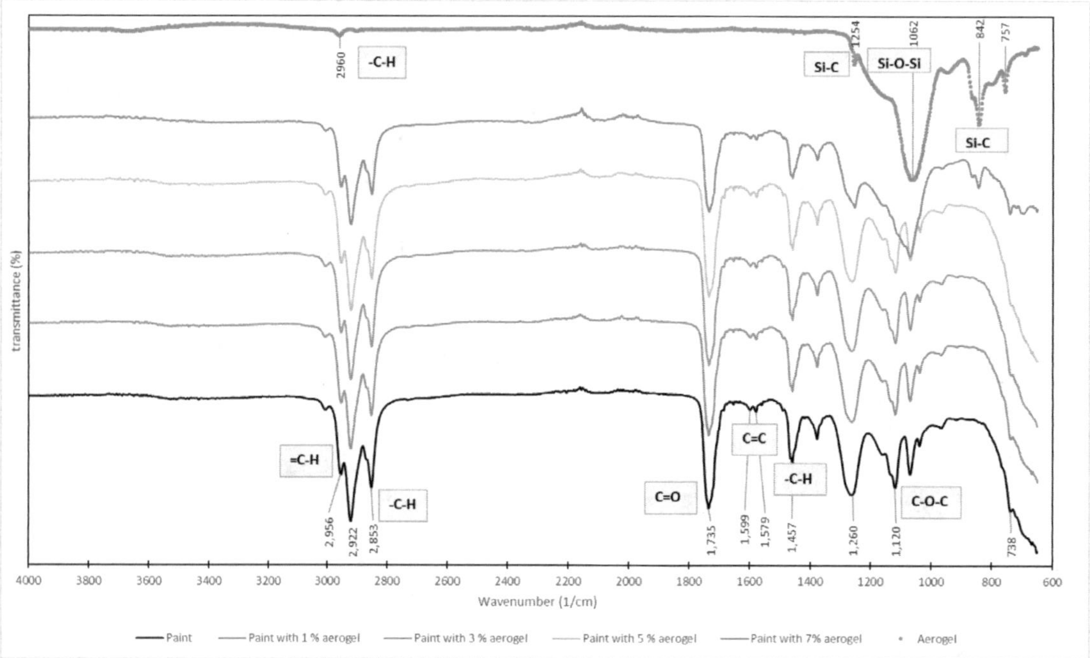

Figure 3. FTIR spectra of paint, paint with 1, 3, 5, and 7 % of Enova silica aerogel content, and Enova silica aerogel.

4. Conclusion

The incorporation of Enova aerogel into the solvent-based paint was conducted to produce thermal insulation paint. In conclusion, the mixture of solvent-based paint and Enova aerogel had been well prepared. The findings can be concluded that the addition of the Enova silica aerogel changed the flow properties of solvent-based paint and the mixture became thicker as the percentage of the Enova aerogel increased. The FTIR spectra revealed that adding 1, 3 and 5% of Enova silica aerogel did not modify the chemical structure and the functional group of the commercial solvent-based paint. However, the addition of 7% Enova silica aerogel revealed that a new peak at 1071 cm^{-1} representing the Si-O-Si bonding had been formed. This condition shows that the Enova silica aerogel was well incorporated into the solvent-based paint. For future studies, the chemical dispersant or any additive such as a wetting agent will be explored to overcome the compatibility and viscosity thickening issue plus tested with heat for the heat insulation study.

5. References

[1] Susila I P, Sukandar I and Santoso B 2020 Real-time acquisition and correction of temperature effect in NaI(Tl) detector-based environmental gamma radiation detection device J. Phys. Conf. Ser. **1436**(1) 012135 doi: 10.1088/1742-6596/1436/1/012135.

[2] Al-Homoud M S 2005 Performance characteristics and practical applications of common building thermal insulation materials *Build Environ.* **40**(3) 353–366 doi: 10.1016/j.buildenv.2004.05.013.

[3] Smirnova I and Gurikov P 2018 Aerogel production: Current status, research directions, and future opportunities *J. of Supercritical Fluids* **134** 228–233 doi: 10.1016/j.supflu.2017.12.037.

[4] Dorcheh A S and Abbasi M H 2008 Silica aerogel; synthesis, properties and characterization doi: 10.1016/j.jmatprotec.2007.10.060.

[5] Smirnova I and Gurikov P 2017 Aerogels in chemical engineering: Strategies toward tailor-made aerogels *Annu. Rev. Chem. Biomol. Eng.* **8**(1) 307–334 doi: 10.1146/annurev-chembioeng-060816-101458.

[6] Baetens R, Jelle B P and Gustavsen A 2011 Aerogel insulation for building applications: A state-of-the-art review doi: 10.1016/j.enbuild.2010.12.012.

[7] He F *et al.* 2019 Thermal Conductivity of Silica Aerogel Thermal Insulation Coatings *Int. J. Thermophys.* **40**(10) doi: 10.1007/s10765-019-2565-6.

[8] Lu Y, Liu Z, Li X, Yin X J and Utomo H Djati 2022 Development of water-based thermal insulation paints using silica aerogel made from incineration bottom ash *Energy Build* **259** doi: 10.1016/j.enbuild.2022.111866.

[9] An L *et al. 2021* Reflective paint consisting of mesoporous silica aerogel and Titania nanoparticles for thermal management *ACS Appl. Nano Mater.* doi: 10.1021/acsanm.1c01280.

[10] Izzo F C, Balliana E, Pinton F, Zendri E 2014 A preliminary study of the composition of commercial oil, acrylic and vinyl paints and their behaviour after accelerated ageing conditions *Conserv. Sci. in Cultural Heritage* **14**(1) 353–369 https://doi.org/10.6092/issn.1973-9494/4753.

[11] De Carvalho R K C, Ortega F S, Morandim-Giannetti A A 2019 Alkyd resin synthesis by enzymatic alcoholysis *Iranian Polym. J.* doi:10.1007/s13726-019-00738-y.

[12] Patil D M, Phalak G A, Mhaske S T 2018 Design and synthesis of bio-based epoxidized alkyd resin for anticorrosive coating application. *Iran Polym. J.* **27** 709–719.

[13] Silverstein R M, Webster F X, Kiemle D J, Bryce D L 2014 Spectrometric identifcation of organic compounds 8th edn. (New York: Wiley).

[14] Nah H Y, Kim Y, Kim T, Lee K Yeon, Parale V G, Lim C H, Seo J Y, Park H H 2020 Comparison studies of surface modification reaction using various silylating agents for silica aerogel J. of Sol-Gel Sci. and Tech. **96** 346–359 https://doi.org/10.1007/s10971-020-05399-5.

[15] Wagh P B, Kumar R, Patel R P, Singh I K, Ingale S V, Gupta S C, Mahadik D B and Rao A V 2015 Hydrophobicity measurement studies of silica aerogels using FTIR spectroscopy, weight difference method, contact angle method and K-F titration method *JCBPS* Section A **5** 2350-2359.

Acknowledgements

The Authors would like to extend their appreciation and sincere thanks to the Malaysian Nuclear Agency including Plant Assessment Technology Group for support and cooperation.

International Nuclear Science Technology and Engineering Conference (iNuSTEC2024) IOP Publishing
IOP Conf. Series: Materials Science and Engineering **1326** (2025) 012005 doi:10.1088/1757-899X/1326/1/012005

Sustaining cyber and ICT security preparedness through disaster recovery testing and simulations

Amy Hamijah Ab. Hamid[1*], Siti Nurbahyah Hamdan[1], Mohamad Safuan Sulaiman[1], Norzalina Nasirudin[1], Mohd. Hasnor Hasan[1], Saa'idi Ismail[1], Mohd Dzul Aiman Aslan[1], Raja Murzaferi Raja Moktar[1], Khairiel Adyani Abd. Ghani[1], Mohd. Amin Abdullah[1], Radhiah Jamalludin[1], Noor Fatin Shuhada Ab. Hamid[a], Hanizan Zainon[a], Mohammad Fazli Rahmat Isa[b], Sufian Norazam Mohamed Aris[1], Mohd Faizul Zakaria[1], Mohamad Rashidi Abd Aziz[1], Siti Farizawaty Sapian[1], Azlinawati Husain[1], Shukri Mohd[1] and Muhammad Rawi Mohamed Zain[1]

[1]Information Technology and Applications Unit (PTM), Technical Support Division, Malaysian Nuclear Agency, Bangi, 43000 Kajang, Selangor, Malaysia.
[a]Radiation Health and Safety Division, Malaysian Nuclear Agency, Bangi, 43000 Kajang, Selangor, Malaysia.
[b]APM Nuclear Technology Sdn. Bhd., Bangi, 43000 Kajang, Malaysia.

[*]amyhamijah@nm.gov.my

Abstract. This paper examines and highlights both challenges and success factors in Cyber and Information Security Incident Management at the Malaysian Nuclear Agency (Nuclear Malaysia) by implementing simulation exercises using a "walkthrough" approach for disaster recovery testing of Information and Communication Technology (ICT). Effective cyber and information security incident management is essential for ensuring compliance and monitoring ICT governance, particularly through the certification of the Cyber and ICT Security Management System (ISMS, ISO/IEC 27001), including the Business Continuity Management System (BCMS, ISO 22301). The goal of these simulations is to evaluate the achievement of disaster recovery objectives and identify potential improvements in ICT disaster recovery strategies at Nuclear Malaysia. The exercises have successfully identified critical achievement levels of PTM incident management, based on Business Impact Analysis (BIA) objectives, through detailed ICT incident management simulation workflows. In conclusion, these simulation exercises verified the BIA objectives achievement levels, ensuring that Nuclear Malaysia's cybersecurity and ICT management are robust and capable of smooth operational normalization. Additionally, they reveal vulnerabilities and weaknesses in current security controls and operational systems, leading to continuous improvements in Nuclear Malaysia's ICT incident management planning.

IOP Conf. Series: Materials Science and Engineering **1326** (2025) 012005 doi:10.1088/1757-899X/1326/1/012005

1. Introduction

Nuclear Malaysia has proactively developed Business Continuity Plan (BCP) and Disaster Recovery Plan (DRP), referring to the International Organization for Standardization (ISO) certification standards for Business Continuity Management System (BCMS, ISO 22301:2012) and Information Technology Security Management System (ISMS, ISO/IEC 27001:2013). Besides that, cyber and information security incident management is a critical requirement in implementing compliance and monitoring of ICT governance according to various governmental circulars, policies, and procedures [1-3][7-11][13]. In regards to these implementations due to strengthening the process for managing scientific software and intangible assets, which is also to sustain efficient and effective ICT management [2][11]. Basically, BCP is a strategy that ensures critical business operations continue during and after a disaster. It is essential for reducing downtime and mitigating risks to business processes. Meanwhile DRP is a documented, structured approach for responding to unplanned incidents that affect IT services. The DRP focuses on restoring IT infrastructure and services.

Nuclear Malaysia is among the pioneering public agencies to be certified in BCMS (ISO 22301:2012) for the provision of services related to critically acclaimed service centres since 2019 [4][8]. These prioritized facilities shall be aided by other supporting divisions such as technology commercialization, engineering, instrumentation and automation services, and information technology support. In addition, the initially certified BCMS is also in competence of the Business Continuity Good Practice Guidelines implementations. In addition, Nuclear Malaysia's Cyber and ICT Security Management System (ISO/IEC 27001:2013) has been initially certified since 2014, focusing on eSSDL System credentials located at our data centre for the provision of data management, web application services, and, in managing eSSDL servers and network management [1][7].

BCMS is a standardized framework designed to ensure an organization's resilience by implementing policies and processes that enable it to maintain operations during and after a disruptive event. BCMS is based on ISO 22301 and focuses on risk assessment, business impact analysis, and recovery planning. In addition, ISMS is a systematic approach to managing sensitive company information to ensure its confidentiality, integrity, and availability. ISMS is based on the ISO/IEC 27001 standard and includes policies, procedures, and controls to mitigate security risks. The exclusion differences between BCMS and ISMS implementations, whereby; BCMS did not include any product development and verification, which includes, critical and not urgent activities and processes. ISMS implementation does not relate to eSSDL customers and business partners.

Previous research has highlighted various challenges in developing the Business Continuity Plans (BCP) and Disaster Recovery Plans (DRP), stemming from inaccurate and inappropriate assumptions about cyber threats due to the following factors [3]:

i. *Lack of expertise:* Expertise shortage among agency staff could contribute to lack of self-reliance of the disaster recovery guidelines in the event of an unexpected cyber incident.
ii. *Differing perspectives and priorities:* Conflict of interest in perspectives and priorities between agency staff and stakeholders of the BCP regarding on how to address the challenges of current cyber threats during disaster plan activation.
iii. *Insufficient reference sources:* The absence of reference sources leads to difficulties in identifying any readiness levels. Insufficient preparedness for recovery actions creates challenges in implementing alternative recovery site operations, whether for local incidents or regional disasters.

IOP Conf. Series: Materials Science and Engineering **1326** (2025) 012005 doi:10.1088/1757-899X/1326/1/012005

2. Analysis methods

Hence, Nuclear Malaysia's BCMS effectiveness validation focuses on how-to achieve organizational resilience in our Business Continuity Strategy Program and Plan (BCSP). Our BCSP is incorporated to identify our strength and effectiveness of the following capabilities as (1) the capability to prevent an accident from occurring; (2) the capability to prevent that an occurred accident spreads its impacts; and (3) the capability to recover to the normal state, after an occurred accident [6]. BCSP is a structured framework aimed at ensuring organizational resilience by integrating risk management, business continuity, and disaster recovery planning. It focuses on preventing incidents, minimizing their impacts, and ensuring rapid recovery. This validation is based on the Business Impact Analysis (BIA) strategy charts for the critically acclaimed processes and activities of the dedicated facilities mentioned above [12]. BIA is a systematic process for evaluating the effects of disruptions on business functions and determining recovery priorities. Meanwhile, the Nuclear Malaysia's ISMS effectiveness validation based on DRP according to the act of maintaining and minimizing system downtime during any disaster and crises that eSSDL specified services must be restored as suggested in Table 1 [1] [18].

Table 1 Critical Dependency of the RPO, RTO and RLO during eSSDL simulation drills [1] [18]

Metric	Definition	Focus	BIA Objectives
RPO (Recovery Point Objective)	Maximum allowable data loss in time	Backup frequency and data protection	• Network Restoration (2 days) • Storage Restoration (3 days) • Application System Full Access (72 hours)
RTO (Recovery Time Objective)	Maximum acceptable downtime before full recovery	System restoration speed	• Network Restoration (4 hours) • Storage Restoration (24 hours) • Application System Full Access (5 days)
RLO (Recovery Level Objective)	Minimum level of service that must be restored within the RTO	Ensuring partial functionality during recovery	• The specified eSSDL services is RLO achievable at least 70% must be operational before the recovery fully completed.

In summary, RPO determines the acceptable data loss, RTO defines the allowable system downtime, and RLO measures the operational capacity during recovery. Establishing these objectives through a Business Impact Analysis (BIA) ensures robust disaster recovery strategies and effective business continuity planning. For example, if the target RPO is 4,320 minutes (72 hours), backups should occur at least every 72 hours to minimize data loss. Similarly, if the target RTO is 7,200 minutes (5 days), the eSSDL system must be restored within a working week. Regardless of meeting the expected RPO and RTO targets, if the RLO is set at 70%, the majority of eSSDL system services must be functional before full recovery is achieved.

The target minutes for RPO (4320 minutes / 72 hours) and RTO (7200 minutes / 5 days) were chosen based on the Business Impact Analysis (BIA) objectives set during the eSSDL simulation drills. These values represent the predefined thresholds for acceptable data loss and recovery time, ensuring minimal disruption to critical operations as the following.

RPO (4320 mins / 72 hours): This is the maximum amount of data loss the organization can tolerate before business operations are significantly impacted. Backups must be performed frequently enough to ensure that, in a worst-case scenario, no more than 72 hours of data is lost.

RTO (7200 mins / 5 days): This represents the maximum acceptable downtime before full recovery must be completed. It considers dependencies on infrastructure, personnel, and system restoration processes to bring operations back online within five days.

These targets were set based on risk assessments, operational requirements, and available recovery resources.

The evaluation of the disaster recovery objectives is according to the RTO restorations achievement during the implementations of the “walkthrough” simulation drill accordingly. Walkthrough simulation drill is a structured exercise where participants systematically review and validate disaster recovery and business continuity plans. This method involves key personnel discussing their roles, responsibilities, and response procedures in a simulated disaster scenario. The primary goals of these drills are:

Evaluate disaster recovery objectives: Assess whether the recovery plans effectively minimize downtime and meet predefined recovery metrics such as RPO, RTO and RLO.

Identify areas for improvement: Detect weaknesses or gaps in the current ICT disaster recovery strategies and refine response measures.
Enhance preparedness: Train employees and stakeholders in crisis response to ensure quick and efficient execution during actual incidents.

Ensure compliance: Validate adherence to industry standards and regulatory requirements, such as ISO/IEC 27001 and ISO 22301.

A critical analysis of the issues related to cybersecurity and ICT management listed above will be discussed in the following section.

3. Results and discussion

Nuclear Malaysia’s Information Technology and Applications Unit (PTM), in partnership with our own Secondary Standard Dosimetry Laboratory (SSDL) from the Radiation Metrology Group (KMS) and APM Technology Sdn. Bhd., has carried out a “walkthrough” simulation drill to evaluate the preparedness and effectiveness of its cyber incident recovery measures. The goal of these exercises is to determine if disaster recovery objectives are being met and to recognize areas for enhancing ICT disaster recovery strategies at Nuclear Malaysia. These simulations were successfully assessed PTM’s incident management performance relative to the proposed Business Impact Analysis (BIA) objectives, as depicted in Figure 1 [14, 15, 16, 17].

The simulation drill "walkthrough" scenario is based on hardware failures at the Bangi Data Centre, which results in access disruption and a malfunction of the database server hardware for the eSSDL System. This drill focused specifically on both BCP and ISMS response as the initial phase of the BIA objectives, which involves identifying the impact of the emergency or incident disruptions on existing processes and services. Essentially, the RTO and Recovery Level Objective (RLO) are determined by evaluating the time requirements (or lead time) and minimum operational needs of the eSSDL system to function effectively following the simulated disaster scenario. The schematic view of the BIA objectives is shown in Figure 2, which suggest that normally the period of the RTO is less than the MTPD because of the time rugs (or lead time) needed for mobilizing both BCP and ISMS responding actions and for preparing resume of the service [5]. MTPD is the longest time an organization can tolerate an operational disruption before significant harm occurs. Meanwhile, time rug is the lead time required to mobilize disaster recovery actions and resume normal operations.

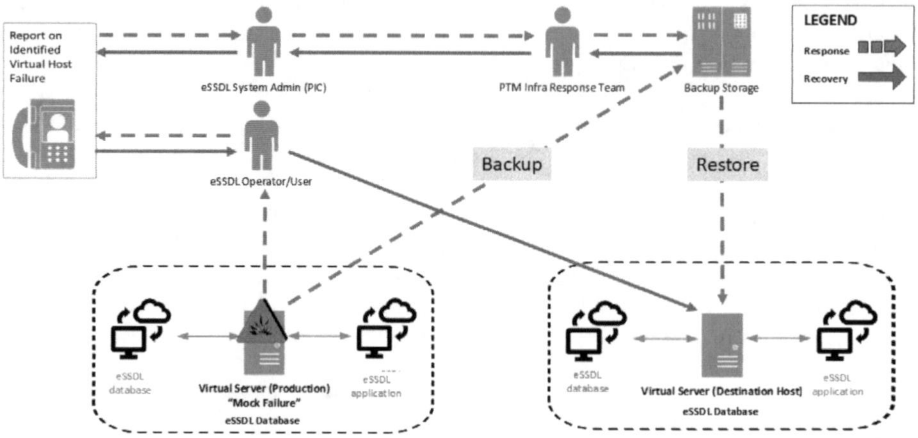

Figure 1. eSSDL server backup and restoration [14, 15, 16, 17]

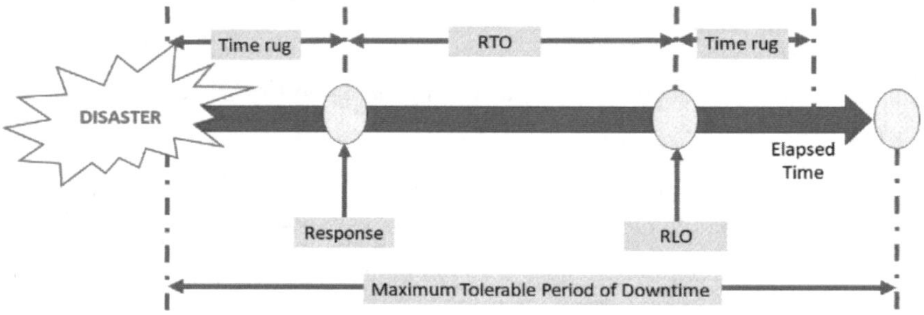

Figure 2. Schematic view of MTPD, RLO and RTO [5]

As a result, Table 2 presents the recent simulation drills and their alignment with the estimated achievements of the BIA objectives as follows.

Table 2 Results comparison during eSSDL simulation drills [14, 15, 16, 17]

Metric	Estimated BIA Objectives (target minutes/percentage)	Results (minutes/percentage)			
		2021	2022	Jan 2024	Nov 2024
RPO	4320 mins (72 hours)	200 mins	372 mins	825 mins	339 mins
RTO	7200 mins (5 days)				
RLO	At least 70% achievable	100%	100%	100%	95%

According to Table 2, to determine the average achievement percentage, we consider the available RLO data: (100 + 100 + 100 + 95) / 4 = 98.75%
Thus, the average achievement for RLO is 98.75%.

Besides that, Table 3 and 4 are the analysed RPO and RTO achievement analyses as suggested.

Table 3 RPO Achievement Analysis

Year	Target (minutes)	Actual Recovery Time (minutes)	Achievement (percentage)
2021	4320 mins (72 hours)	200 mins	4.63%
2022	4320 mins (72 hours)	372 mins	8.61%
Jan 2024	4320 mins (72 hours)	825 mins	19.10%
Nov 2024	4320 mins (72 hours)	339 mins	7.85%
Average RPO Achievement: (4.63 + 8.61 + 19.10 + 7.85) ÷ 4 = 10.05%			

Table 4 RTO Achievement Analysis

Year	Target (minutes)	Actual Recovery Time (minutes)	Achievement (percentage)
2021	7200 mins (5 days)	200 mins	2.78%
2022	7200 mins (5 days)	372 mins	5.17%
Jan 2024	7200 mins (5 days)	825 mins	11.46%
Nov 2024	7200 mins (5 days)	339 mins	4.71%
Average RTO Achievement: (2.78 + 5.17 + 11.46 + 4.71) ÷ 4 = 6.03%			

Therefore, the average achievements of the eSSDL simulation drills in relation to the BIA objectives are as follows:

- RPO Achievement: 10.05% estimated minimized data loss.
- RTO Achievement: 6.03% acceptable downtime.
- RLO Achievement: 98.75% system services restoration.

These results highlight that while RLO has been effectively met, further improvements in data recovery (RPO) and system downtime reduction (RTO) are necessary to enhance disaster recovery effectiveness.

Despite having a structured disaster recovery framework, various factors may cause delays and inefficiencies in meeting BIA objectives, including:

Uncertainty in Resolving Incidents: Some cyber incidents, like ransomware attacks or hardware failures, necessitate thorough forensic analysis and mitigation efforts before restoration can commence.

Infrastructure Limitations: Dependency on aging or legacy infrastructure may result in longer lead times for hardware replacement or system reconfiguration.

Inadequate Testing and Drills: If disaster recovery simulations are not conducted regularly or thoroughly, organizations may underestimate the actual recovery time needed.

Resource Constraints: Limited human resources and skill gaps in cyber incident response teams may slow down execution during a real disaster scenario.

Synchronization Challenges: Restoring large-scale data, particularly in distributed environments, demands considerable time for validation and consistency checks before services can be resumed.

Network Bottlenecks and Connectivity Challenges: Poor network performance or failure in redundancy mechanisms can significantly impact recovery timelines.

4. Conclusions

In conclusion, these simulation exercises have enhanced the validation of DRP and BIA objectives, ensuring that Nuclear Malaysia's cybersecurity and ICT management are effectively maintained through the smooth normalization of ICT governance and operations. Despite successes in meeting RTO targets, challenges remain in sustaining a consistent 24-hour recovery time. Addressing factors such as resource availability, dependency on third-party vendors, and network limitations will be critical for improving response capabilities. Consequently, PTM's cybersecurity and ICT incident management planning at Nuclear Malaysia can be continuously refined and improved. Looking ahead, our team plans to advance these simulation exercises by integrating a digitization module for large-scale consolidation within Nuclear Malaysia. As to integrate digital automation to streamline disaster recovery processes and enhance efficiency across multiple departments.

References

[1] Agensi Nuklear Malaysia 2023 *Dokumen ISMS.* https://sp-kms.nuclearmalaysia.gov.my/teknikal/bst/itc/isms/SitePages/DokumenISMS.aspx

[2] Agensi Nuklear Malaysia 2023 Pelan Strategik Pendigitalan Agensi Nuklear Malaysia: Pembentangan Dokumen Pelan *Bengkel Pemurnian dan Pengesahan Dokumen Pelan Strategik Pendigitalan Agensi Nuklear Malaysia 2012 – 2025, Institut Latihan Islam Malaysia (ILIM), Bandar Baru Bangi, Selangor, Malaysia.*

[3] Hamid A H A 2018 Limitations and challenges towards an effective business continuity management in Nuklear Malaysia *IOP Conf, Ser.: Mater. Sci, and Eng.* **298**(012050) 11.

[4] Hamid A H A, Rozan M Z A, Ibrahim R, Deris S, Yunus M N M 2014 Framing a Nuclear Emergency Plan using Qualitative Regression Analysis. *J. of Nucl. and Rel. Tech.* (JNRT) **11**(1) 27-45.

[5] Caselli F, Gallardo M R, Beale M, Akakura Y and Ono K2016 Methodology and procedure of business impact analysis for improving port logistics business continuity management. *IDRiM J.* **6**(1).

[6] Giacchero A, Giordano F and Schiraldi M 2013 From business continuity to design of critical infrastructures: ensuring the proper resilience level to data centres. **Int. J. of Eng. and Tech. (IJET) 5**(4) 3544-3553.

[7] International Organization for Standardization 2013 Information technology - Security techniques - Information security management systems - Requirements ISO/IEC 27001Second Ed. -10-01.

[8] International Organization for Standardization 2019 Security and resilience - Business continuity management systems - Requirements ISO 22301 Second Ed. 10.

[9] Kerajaan Malaysia (Semakan dan Pindaan 2017) *Arahan Keselamatan.*

[10] Kerajaan Malaysia (Semakan dan Pindaan 2022) Pekeliling *Am Bilangan 4 Tahun 2022*: *Pengurusan dan Pengendalian Insiden Keselamatan Siber Sektor Awam.*

[11] Sulaiman M S B, Mohd S B, Hamdan N et al. 2023 Penyediaan SOP bagi Memperbaiki Mutu Perbincangan Teknikal di dalam Mesyuarat JPICT Agensi Nuklear Malaysia. *Konvensyen Teknikal dan Inovasi Agensi Nuklear Malaysia (NITC2023) (*Agensi Nuklear Malaysia)

[12] Zalil N F B 2019 Predictive Analytics: Analysing Effectiveness of the Business Continuity Recovery Response Simulation System (BCRRSS) in Nuklear Malaysia *Practicum Report (*Universiti Utara Malaysia).

[13] Hamdan S N B, Sulaiman M S B, Ibrahim M B *et al.* 2023Kepentingan penyelenggaraan sistem maklumat dalam kitaran hayat pembangunan sistem (SDLC) dan pelan kesinambungan perkhidmatan (PKP) Agensi *Konvensyen Teknikal dan Inovasi Agensi Nuklear Malaysia (NITC2023)* (Malaysia: Agensi Nuklear Malaysia).

[14] Unit Teknologi Maklumat dan Aplikasi 2021 *Laporan Pengujian Pemulihan Bencana ICT eSSDL* (Kajang: Agensi Nuklear Malaysia).

[15] Unit Teknologi Maklumat dan Aplikasi 2022 *Laporan Pengujian Pemulihan Bencana ICT eSSDL* (Kajang: Agensi Nuklear Malaysia).

[16] Unit Teknologi Maklumat dan Aplikasi 26 Januari 2024 *Laporan Pengujian Pemulihan Bencana ICT eSSDL* (Kajang: Agensi Nuklear Malaysia).

[17] Unit Teknologi Maklumat dan Aplikasi 26 November 2024 *Laporan Pengujian Pemulihan Bencana ICT eSSDL* (Kajang: Agensi Nuklear Malaysia).

[18] Unit Teknologi Maklumat dan Aplikasi 2024 *Pelan Pemulihan Bencana ICT versi 6.0* (Kajang: Agensi Nuklear Malaysia).

Acknowledgement

The authors would like to extend their sincere gratitude to the eSSDL system simulation drill response team from Nuclear Malaysia and APM Nuclear Technology Sdn. Bhd. for their exceptional teamwork and expertise. We hope that the ongoing collaboration between SSDL, PTM, and APM Nuclear Technology Sdn. Bhd. will continue to positively influence and inspire progress in the field of ICT and, nuclear safety and security research.

Phosphate removal and yttrium recovery from xenotime mineral via alkaline fusion

Roshasnorlyza Hazan, Nurhusnina Akma Kamarudin, Nurhanisah Asri, Nur Aqilah Sapiee, Khaironie Mohamed Takip, Wilfred Paulus, Norhazirah Azhar and Jacqueline Kones

Materials Technology Group, Industrial Technology Division, Malaysian Nuclear Agency, Bangi, 43000, Kajang, Selangor.

roshasnorlyza@nm.gov.my

Abstract. In Malaysia, heavy rare earth elements (HREE) can be obtained from xenotime minerals. This study investigated the process of phosphate removal from xenotime through alkaline fusion, which involves fusing the minerals with sodium hydroxide (NaOH) pellets. In order to crack the mineral, the xenotime mineral was fused with NaOH in a 1:2 ratio at 350 °C for 3 hours. The next step involved washing the fused mixture with hot deionized water (~80 °C) 80 °C) until the pH of the fused mixtures was neutral (pH 6 to pH 8), followed by agitation and filtering steps. The fused mixture and its filtrate were analysed using X-ray fluorescence (XRF) for elemental analysis. Then, the solid sample undergoes an acid leaching process to elucidate the effect of the washing method on the fused xenotime leaching process. These findings collectively highlight alkaline fusion approaches to efficiently remove phosphate (100%) from xenotime minerals involving hot water rinses.

1. Introduction

Xenotime is a rare earth phosphate mineral that contains significant amounts of yttrium and other heavy rare earth elements. The efficient removal of phosphate from xenotime is crucial for the subsequent processing and purification of the rare earth elements [1]. Earlier research has shown that phosphoric acid leaching can be an effective method for extracting rare earth elements from phosphate-containing minerals. However, the residual phosphate in the solution can still pose challenges in the downstream processing [2-3].

Rare earth elements have become increasingly critical for various high-tech applications, including renewable energy technologies [4]. One such element, yttrium, is widely used in compact fluorescent lamps and wind turbines, making its supply chain vulnerable to environmental and geopolitical concerns [4-5]. Xenotime, a rare earth phosphate mineral, is a promising secondary source of yttrium, but its extraction and processing present unique challenges. Recognizing the need for more sustainable and efficient utilization of this critical raw material, intensive study to investigate the potential of using alkaline fusion as a means to simultaneously remove phosphates and recover yttrium from xenotime.

The extraction and separation of rare earth elements from their ores is a complex process, often requiring intensive chemical treatments and generating significant environmental impacts [6]. Existing literature on the beneficiation of rare earth minerals has primarily focused on the two major deposits,

Bayan Obo in China and Mountain Pass in the USA, leaving a gap in the understanding of less prominent but equally important deposits, such as xenotime. Pyrometallurgical approaches, including molten salt electrolysis, have shown promise in extracting and purifying specific rare earth elements, such as neodymium, but their applicability to yttrium recovery from xenotime remains largely unexplored [7]. By leveraging the differential solubility of rare earth phosphates and oxides in alkaline media, the proposed approach aims to simultaneously purify the rare earth fraction and facilitate the recovery of yttrium as a valuable co-product.

Processing minerals like xenotime to extract valuable elements like Y and rare earth metals is known as alkaline fusion [8-9]. The alkaline fusion process involves the high-temperature treatment of xenotime mineral with a basic reagent, such as NaOH or sodium carbonate [10]. Alkaline fusion decomposes phosphate-containing minerals, such as xenotime, into hydroxides. This will facilitate subsequent hydrometallurgical processes in phosphate removal [11]. This process converts the phosphate in the xenotime mineral into a form that can dissolve in water [12]. This makes it simple to separate the phosphate from the rare earth-containing residue. Additionally, alkaline fusion has other advantages, such as being environmentally friendly, and this method is an alternative process to replace hazardous (toxic) concentrated sulphuric acid to crack minerals.

The key steps in the alkaline fusion process are as follows: First, the xenotime mineral is mixed with an alkaline reagent like NaOH or sodium carbonate [10, 13]; then, the mineral-reagent mixture is fused at high temperature; then, water is leached through the fused product to dissolve the phosphate; and finally, the residue containing rare earths is filtered out of the solution rich in phosphate.

The optimal parameters for the alkaline fusion process, including the reagent type, reagent-to-mineral ratio, temperature, and fusion time, need to be thoroughly investigated to achieve complete phosphate removal and maximise the recovery of rare earth elements [14]. An article from 2023 says that the alkaline fusion process could be used to remove phosphate from apatite and monazite, among other minerals that contain phosphate [10]. However, the presence of phosphate in xenotime minerals hindered REE recovery by disrupting acid leaching efficiency. Therefore, additional research is necessary to enhance the phosphate removal process.

Phosphate removal from xenotime mineral resources is essential to minimise environmental impacts and maximise the recovery of valuable rare earth elements. The presence of phosphate in mineral processing can lead to the formation of undesirable precipitates, which can hinder the overall efficiency of the extraction process. Researchers have explored conventional methods like chemical precipitation and biological uptake to remove phosphate from wastewater, but their applicability to mineral processing remains limited [15]. Researchers have also explored electrochemical methods for phosphate removal from wastewater, showcasing their potential for low-energy consumption and selective precipitation of calcium phosphate [16]. However, no extensive studies have been conducted on the application of these approaches to xenotime mineral processing. In this research paper, we investigate the use of an alkaline fusion process to achieve total phosphate removal from xenotime minerals.

2. Methodology

2.1. Materials

The xenotime ores were collected from Kinta Valley, Perak, Malaysia. Table 1 displays the major elements present in the xenotime ore. Other chemicals, such as NaOH, 99% in pellet form manufactured by R&M Chemical, and hydrochloride acid (HCl), 37% manufactured by Merck, were analytical grade.

IOP Conf. Series: Materials Science and Engineering **1326** (2025) 012006 doi:10.1088/1757-899X/1326/1/012006

Table 1. Elemental composition of xenotime.

Element	Elemental composition (wt. %)
Rare Earth elements (REE)	
Y	21.98
Gd	1.374
Tm	ND
Ho	0.66
Yb	5.57
Nd	1.96
Ce	1.79
Radioactive elements	
Th	0.45
U	0.38
Inorganic elements	
P	41.88
Other elements	23.95

*ND = not detected

2.2. Experimental

Mineral cracking was done by mixing xenotime minerals with NaOH beads and fused in the furnace at 300 °C for 3 hours. This study investigated five different series of experimental washing conditions. Table 2 displays the various experimental conditions under investigation. In a typical alkaline fusion experiment was labelled as Experiment Series 2. In order to see the effect of temperature and agitation on the fused sample, Series 1 was left for 3 hours without any agitation at room temperature. The mixture's final pH remains at 14. After washing and filtration, the solid sample undergoes an acid leaching process.

Table 2. Experimental conditions.

Parameters	Experiment Series 1	Experiment Series 2	Experiment Series 3	Experiment Series 4	Experiment Series 5
Water leaching temperature (°C)	25	25	25	80	80
Water leaching agitation	No	Vigorous stir for 3 hours	Vigorous stir for 3 hours	Vigorous stir for 3 hours	Vigorous stir for 3 hours
Rinsing after filtration steps	No	No	80 °C distilled water	No	80 °C distilled water
Rinsing water volume (ml)	No	No	2000	No	1775
Total volume (ml)	900	900	2900	900	2675
Final mixture pH	14	14	8	14	7

2.3. Acid leaching performance for verification.

The effect of phosphate removal on acid leaching efficiency was examined. To understand the impact of the washing method on the fused xenotime leaching process, the leaching was conducted in an air atmosphere at 70-80 °C for six hours. This step involved the introduction of 6 M HCl. The solution was quickly separated by vacuum filtration after the acid leaching step.

2.4. Analytical techniques
Shidmadzu's EDXRF 7000 performed the elemental analysis for both liquid and solid samples. For morphology of the samples, Field Emission Scanning Electron Microscope (FESEM), Carl Zeiss, GeminiSEM 500 equipped with Energy Dispersive X-Ray Spectrometer (EDXs), X-Max 80 by Oxford Instrument were used. Three points were selected to pinpoint the presence of the element in a particular morphology.

2.5. Evaluation of data

The element recovery (R%) [13] is calculated as follows:

$$R\% = \frac{weight\ of\ an\ element\ after\ alkaline\ fusion}{weight\ of\ an\ element\ in\ xenotime\ mineral} \times 100 \quad (1)$$

3. Results and discussion
Table 3 shows that the amounts of most of the elements in the fused mixture of xenotime were clearly different between the experiment series. The phosphorus (P) content was one of the most noticeable differences. In experiment series 1, there is still 7.35% of P. Even after vigorously stirring for 3 hours in experimental series 2, there is still P present (4.92%) in the fused mixture.

Table 3. Elemental composition of xenotime before and after alkaline fusion process.

Element	Elemental composition (wt. %)				
	Experiment Series 1	Experiment Series 2	Experiment Series 3	Experiment Series 4	Experiment Series 5
Rare Earth elements (REE)					
Y	47.31	51.52	52.72	49.52	50.73
Gd	2.93	2.83	3.06	3.17	3.26
Tm	1.63	3.88	2.07	5.09	4.84
Ho	ND	2.01	2.4	1.43	2.15
Yb	12.71	7.96	9.32	9.50	8.81
Nd	4.29	4.35	3.00	4.76	5.14
Ce	3.68	2.99	2.30	4.18	4.61
Radioactive elements					
Th	1.07	1.14	0.92	1.11	1.14
U	0.75	0.79	0.91	0.78	0.78
Inorganic elements					
P	7.35	4.92	ND	ND	ND
Other elements	18.28	17.62	23.29	20.48	18.54

*ND = not detected

Therefore, experimental series 4 was planned to study the effect of vigorously stirring with hot water (80 °C) for 3 hours. This series successfully removes P from the fused mixture. However, the fused mixture's pH is still very high (pH 14). This condition needs to be prevented, as in the next steps, acid leaching will be introduced to the fused mixture. Therefore, acids and alkaline-fused mixtures react with each other to neutralise each other's properties first rather than leach out REE in the fused mixture. The transfer of protons (H^+ ions) from the acid to the alkaline fused mixture drives this reaction. To neutralise the mixture, experimental series 5 was conducted, and the result also shown in Table 2. Although the composition did not significantly differ, Table 2 states that the final mixture's pH value ultimately dropped to 7. Additionally, agitation in hot water reduces the usage of distilled water from 2000 ml to 1775 ml during the rinsing step. Even though the difference in total volume was too small, it cannot be neglected as it provides useful information if this laboratory scale needs to be scaled up in the future. The design of the demo and pilot plant should consider this outcome. Even though the differences are quite significant, the presence of P in the mixture will affect the subsequence step in the purification process for REE extraction.

Figure 1 illustrates the 100% recovery of Y elements across all experiment series. These were also applied to P removal (%) in all of the experiment series, except for series 1 and 2. Sample series 1 and 2 still had 27.1% and 19.9% P remaining, respectively.

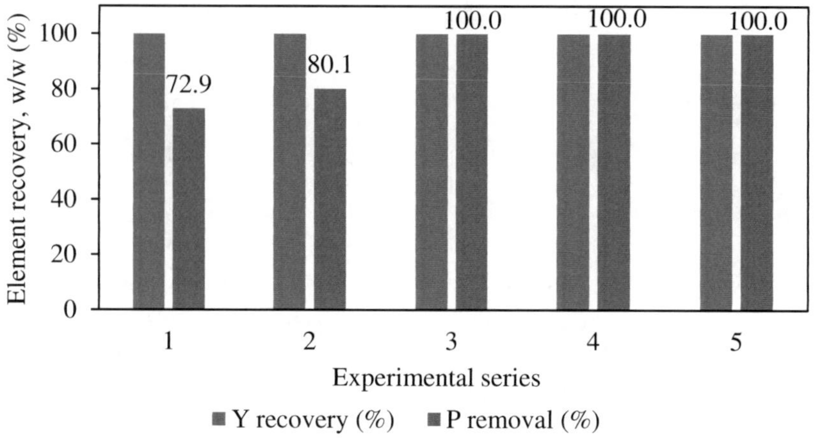

Figure 1. Y and P recovery after alkaline fusion process.

When the xenotime mineral or its chemical formula $REPO_4$ or YPO_4 mixes with NaOH, the following reactions occur: Equation (2) and Equation (3) [18]. $REPO_4$ represents the mineral's rare earth content, whereas YPO_4 represents specifically the Y element in the xenotime mineral.

$$2REPO_4 + 6NaOH \rightarrow 2RE(OH)_3 + 2Na_3PO_4 \quad (2)$$

$$2YPO_4 + 6NaOH \rightarrow 2Y(OH)_3 + 2Na_3PO_4 \quad (3)$$

Figure 2a and 2b show FESEM micrographs of the fused samples from experimental series 2 and experimental series 5 respectively. Fused sample from series 2 displays the fibre structure which consists mainly of P elements (Table 4).

Thus, based on the analytical analysis, it can be concluded that water leach temperature and agitation play important roles in removing P from the xenotime mixture as proven for Experiment Series 5 as compare to typical alkaline fusion method (Experiment Series 2) and shown in Table 4. It is important to rinse the mixture with hot water, approximately 80 °C, to critically remove the P from the xenotime mineral. In addition, water usage also can be reduce.

The leaching performance was conducted to investigate whether the presence of P in the fused samples could result in a deficiency during the leaching process. Table 5 displays the major elements present in the leachate. Xenotime concentrate from the fused mixture was solubilised. The leachate from the fused mixture without P contains 29.76% Y, while the leachate from the fused mixture with P only contains 4.84% Y.

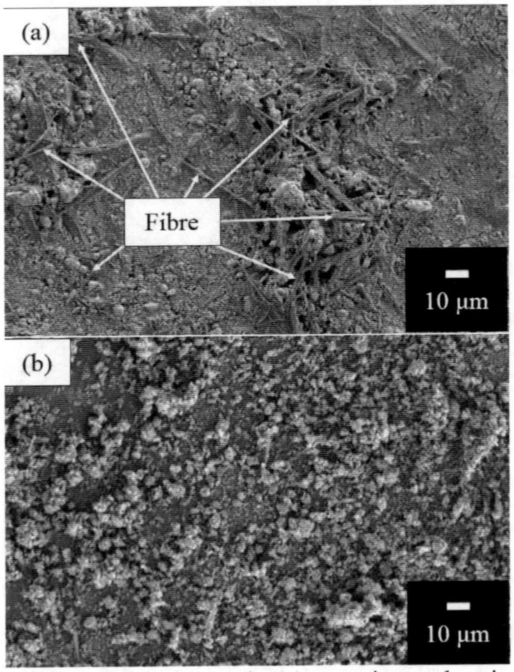

Figure 2. FESEM micrographs of fused sample (a) experimental series 2 and (b) experimental series 5.

Table 4. Element composition / EDAX results of fused samples in Figure 2.

Element	Elemental composition (wt. %)	
	Experiment Series 2 (Fibre morphology)	Experiment Series 5
Y	23.25	36.30
Dy	1.66	5.80
P	44.31	0
Other	30.78	57.90

There were huge differences in content between leachates, showing that the existence of P could hinder the Y liberation from the xenotime-fused mixture. The leachate exhibits an increase in the content of other REEs, with the exception of dysprosium (Dy) and cerium (Ce). This is because the leaching properties of each element are unique and may differ from those of other elements. Prameswara et al. proposed in 2023 that the characteristics of the ore, mass transfer dynamics, and reaction kinetics typically govern the leaching process [17]. Equations 4 describe the standard process for dissolving rare earth element hydroxide in an HCl solution, with a special focus on Y hydroxide in Equations 5 [19]. Furthermore, the high chlorine concentration in both samples indicates that the leachate contains unreacted hydrochloric acid.

$$RE\,(OH)_3 + 3HCl \rightarrow RECl_3\,(aq) + 3H_2O \quad (4)$$
$$Y\,(OH)_3 + 3HCl \rightarrow YCl_3\,(aq) + 3H_2O \quad (5)$$

Table 5. Elemental composition of leachate after acid leaching process.

Element	Elemental composition (wt. %) in leachate	
	Without P (Experiment Series 5)	With P (Experiment Series 2)
Rare Earth elements (REE)		
Y	29.76	4.84
Dy	ND	1.37
Yb	3.89	0.75
Ce	ND	0.56
Tm	1.84	0.44
Nd	0.78	0.38
Gd	1.28	0.31
La	0.52	0.28
Ho	0.92	0.24
Sm	0.25	ND
Radioactive elements		
Th	ND	0.13
U	0.42	0.08
Inorganic elements		
Cl	52.88	88.60
Other elements	7.46	2.02

*ND = not detected

After alkaline fusion of xenotime, the yttrium and other REEs are present as rare earth hydroxides [20]. The alkaline fusion process breaks down phosphate bonds and silica matrices in the minerals, making them more susceptible to acid leaching [20]. These compounds have limited solubility in acidic solutions. Directly adding acid to the highly alkaline fusion product would not efficiently dissolve the REEs. The neutralization step, by slowly adding acid, brings the pH down to a level where the REEs can react with the acid and dissolve. This is essential for the subsequent acid leaching step, where the goal is to dissolve the REEs into solution for further processing and separation. Without the neutralization step, much of the yttrium and other REEs would remain in the solid phase and not be recovered.

4. Conclusion

The removal of phosphate from xenotime minerals is essential for the efficient extraction and purification of rare earth elements. The alkaline fusion process presents a viable approach to achieve total phosphate removal, but the process parameters need to be carefully investigated and optimised for xenotime minerals. Therefore, to ensure complete phosphate removal, the xenotime-fused mixture was stirred with 80 °C water and rinsed with hot water until the pH is neutral (pH 7-8). The leachate from the fused mixture without P recovered 29.76% of Y, while the leachate from the fused mixture with P only recovered 4.84% Y.

References

[1] Yang X, Makkonen H V and Pakkanen L 2019 *Minerals* **9** 262
[2] Laurino J P, Mustacato J and Huba Z J 2019 *Minerals* **9** 477
[3] Soukeur A, Szymczyk A, Berbar Y and Amara M 2021*Separation and Purification Technology* **256** 117857
[4] Zhang K., Kleit A N and Nieto A 2017 *Renewable and Sustainable Energy Reviews* **77** 899.
[5] Favot M and Massarutto A 2019 *Journal of Environmental Management* **240** 504.
[6] McLellan B, Corder G and Ali S H 2013 *Minerals* **3** 304.
[7] Sarfo P, Das A and Young C 2020 *Separation and Purification Technology* **256** 117770.
[8] Prameswara G, Trisnawati I, Poernomo H, Mulyono P, Prasetya A and Petrus H T M B 2020 *Mining Metallurgy & Exploration* **37** 1297
[9] Azhar N, Mohamed Takip K, Kones J, Sapiee N A and Zahari A K 2019 *Jurnal Sains Nuklear Malaysia* **31** 11
[10] Merzeg F A, Bezzi N, Bouzidi N, Narsis S, Bait N, Ladji R and Bachari K 2023 *Solid State Physics Mineral Processing* **1** 61
[11] Zhang M, Zhang H, Xu D, Han L, Zhang J, Zhang L, Wu W and Tian B 2011 *Separation Science and Technology* **46** 2260
[12] Luo X, Zhang Y, Zhou H, He K, Luo C, Liu Z and Tang X 2022 *Minerals* **12** 554.
[13] Julapong P, Numprasanthai A, Tangwattananukul L, Juntarasakul O, Srichonphaisarn P, Aikawa K, Park I, Ito M, Tabelin C B and Phengsaart T 2023 *Appl. Sci.* **13** 8364.
[14] Wu S, Wang L, Zhao L, Zhang P, El-Shall H, Moudgil B, Huang X and Zhang L 2018 *Chemical Engineering Journal* **335** 774
[15] Parsons S A and Smith J A 2008 *Elements* **4** 109–112.
[16] Lei Y, Geraets E, Saakes M, Weijden R D and Buisman C J N 2020 *Water Research* **169** 115207
[17] Prameswara G, Trisnawati I, Handini T, Poernomo H, Mulyono P, Prasetya A and Petrus H T M B 2023 *International Journal of Technology* **14** 770
[18] Steyadji M and Nurly H F 2018 *J. Phys.: Conf. Ser.* **1091** 012012
[19] Samin, Setiawan K, Anggraini M and Sunanti S T 2020 *J. Phys.: Conf. Ser.* **1436** 012070
[20] Trisnawati I, Prameswara G, Mulyono P, Prasetya A and Petrus H T B M 2020 *International Journal of Technology* **11** 804.

Acknowledgments

The authors really appreciate the financially supported from MNS-XAPP Malaysia.

International Nuclear Science Technology and Engineering Conference (iNuSTEC2024) IOP Publishing
IOP Conf. Series: Materials Science and Engineering **1326** (2025) 012007 doi:10.1088/1757-899X/1326/1/012007

Influencing factors for spent nuclear fuel from research reactor on radioactive waste management plan

Nurul Muyassarah Ibrahim[1], **Syazwani Mohd Fadzil**[1,2*], **Nurul Syakirah Zainal**[1], **Rohyiza Ba'an**[3], **Muhammad Khairul Ariff Mustafa**[4]

[1]Department of Applied Physics, Faculty of Science and Technology, Universiti Kebangsaan Malaysia, 43600 Bangi, Selangor, Malaysia

[2]Nuclear Technology Research Centre, Faculty of Science and Technology, Universiti Kebangsaan Malaysia, 43600 Bangi, Selangor, Malaysia

[3]Waste and Environmental Technology Division, Malaysian Nuclear Agency, 43000 Kajang, Selangor, Malaysia

[4]Reactor Technology Centre, Malaysian Nuclear Agency, Bangi, 43000, Kajang, Selangor, Malaysia

*syazwanimf@ukm.edu.my

Abstract. Research reactors have various functions in the development of nuclear technology, especially for countries that support nuclear energy for peaceful purposes. Much of the high-level spent nuclear fuel generated from research reactors currently does not have a good disposal plan. This has caused debate from the public. However, ongoing efforts in research and development in producing more strategic management of spent nuclear fuel (SNF) to help nuclear power and the environment are growing rapidly. SNF management is defined as the efficient and safe handling, storage and disposal of SNF, and radioactive waste generated by nuclear power plants or other nuclear reactors. This study aims to look in detail at the management method of spent nuclear fuel from research reactors and identify the factors that influence the disposal plan in Malaysia. In addition, this research also uses a qualitative method where a combination of two approaches was used, which is an analysis of international and national legal documents and a focused cross-check method with related documents. Therefore, a review of the strategy for the disposal of spent nuclear fuel from research reactors in Malaysia should be emphasised to ensure that the country's radioactive waste management is more strategic and optimal in order to increase confidence and subsequently change the public's perception of nuclear energy.

1. Introduction

Nuclear energy often causes heated debates as an alternative source of energy due to its social acceptance, or lack thereof across the world regarding the management of radioactive wastes. Recently, the issue has caused increased resistance from the public in multiple countries, and hence radioactive waste management has faced even more challenges. Detractors that oppose nuclear energy are influenced by several factors, including safety, risk of nuclear proliferation, high labour costs and radioactive waste management. This highlights the fact that plans for spent nuclear fuel (SNF) management have yet to be sufficiently robust to convince the masses in its efficacy, showing that detailed analysis into radioactive waste management plans are required. SNF management is defined as the efficient and safe handling, storage and disposal of SNF, and radioactive waste generated by nuclear power plants or other nuclear reactors.

Malaysia is categorised as a Non-Nuclear Weapon State (NNWS) which does not directly involve the use of nuclear weapons, however nuclear energy is utilised via a research reactor [1]. Thus, peaceful applications of energy from nuclear reaction have become a cornerstone to Malaysian policy, and alongside the development of science and technology, Malaysia has built a research reactor located in the Malaysian Nuclear Agency (ANM) in Dengkil, Selangor, which commenced its operation in 1982 as the sole research reactor in Malaysia [2]. However, SNF management in Malaysia has yet to be planned properly and is currently in the discussion phase. As SNF is categorised as high-level radioactive waste, it is imperative to study the management method of spent nuclear fuel from research reactors and identify the factors that influence the disposal plan in Malaysia. In short, SNF disposal is a complex and continuous process, involving various choices for storage, transport and long-term disposal to ensure a safe and secure management for high-level radioactive wastes.

2. Methodology

2.1. Study design

A qualitative method is used for this particular study, where observations and document analysis are used as a means to express findings, as opposed to a quantitative method involving numbers and values. On the other hand, further elaboration on quantitative data in the form of text enables a more thorough understanding of studied phenomena [3]. In this context, the choice to employ a qualitative study stem from the need to survey various SNF disposal plans for research reactors in order to obtain detailed plans regarding the relevant waste disposals.

Several methods will be used to interpret the results, however this study only focuses on legislation document analysis via cross-checking international and national legislation documents, as well as data from written documents, such as past and international studies.

2.2. Data gathering

All sources concerning radioactive waste management have been studied in detail to extract the relevant information for this study. Amongst the various mediums used include books, journals, articles, seminars and legislation documents. Multiple journals have also been acquired via online sources, such as Google Scholar, Academic.edu and UKM e-journal. The documents are then properly reviewed for the relevant information, specifically factors on the management of SNF in radioactive waste management plans.

Software such as Atlas.ti is also used for literature review on international and national documents. This involves checking for relevant keywords within the requirements in Malaysia and in several countries to determine which factors are prioritised. The software also observes the similarities of the international document that have been fulfilled by Malaysia to ensure the factors are properly accounted for in both documents.

The cross-checking method is very crucial, where the factors contained within the international legislation are checked against what is currently present in Malaysia. The comparisons also involve technical manuals and safety guidelines regarding disposal methods in addition to suitable scenarios for radioactive waste management in Malaysia. The international manual will serve as an example for the influencing factors for spent nuclear fuel management that have to be emphasised for the development of a plan suitable for the laws of Malaysia.

2.3. Data analysis

Data analysis is carried out manually without any software. The analysis is carried out with a cross-sectional technique, where multiple aspects of the study are analysed according to its relevance and relationship to one another. Amongst the reading materials that are checked and analysed include international legislation and guidelines, national legislation documents, and radioactive material management plans from other countries as listed in Table 1.

Table 1 List of references that used during cross checking analysis

No.	Documents	References
1.	General Safety Requirements GSR Part 3, Radiation Protection and Safety of Radiation Sources: International Basic Safety Standards	[4]
2.	General Safety Requirements GSR Part 5, Predisposal Management of Radioactive Waste	[5]
3.	Specific Safety Requirement SSR-4, Safety of Nuclear Fuel Cycle Facilities	[6]
4.	Specific Safety Requirement SSR-5, Disposal of Radioactive Waste	[7]
5.	Specific Safety Guide SSG-1 (Rev. 1), Borehole Disposal Facilities for Disused Sealed Radioactive Sources	[8]
6.	Specific Safety Guide SSG-14, Geological Disposal Facilities for Radioactive Waste	[9]
7.	Specific Safety Guide SSG-15, Storage of Spent Nuclear Fuel	[10]
8.	Specific Safety Guide SSG-42, Safety of Nuclear Fuel Reprocessing Facilities	[11]
9.	Policies and Strategy for Radioactive Waste Management, Nuclear energy Series No. NW-G-1.1	[12]
10.	Option for Management of Spent Nuclear Fuel and Radioactive Waste for Countries Developing New Nuclear Power Programmes, Nuclear Energy Series No. NW-T-1.24	[13]
11.	Storing Spent Fuel until Transport to Reprocessing or Disposal, Nuclear Energy Series No. NF-T-3.3	[14]
12.	Available Reprocessing and Recycling Services for Research Reactor Spent Nuclear Fuel , Nuclear Energy Series No. NW-T-1.11	[15]

The findings are then analysed by cross checking the documents and highlighting two factors for further discussion. The analysis will be finally brought forward during the interview session in future study with referred experts who will then select the optimal disposal plan for further development in Malaysia.

3. Results and discussion

This study has taken into international and national documents to identify the factors influencing radioactive waste management plan on SNF. Previous literature explaining the complexity of storing highly active radioactive waste requires special, highly secure facilities and is expensive; facilities for disposing of these wastes have not been available and it has not been possible to develop them; there has been considerable public opposition to siting radioactive waste facilities away from the sites where wastes are stored [16]. Based on listed documents in Table 1, several factors have been chosen as important factors that influence and act as a starting framework for the development of a SNF disposal plan for the research reactor in Malaysia. The two factors that were chosen here are technical considerations, as well as society and politics. Future work will focus more on economic implications for SNF management. Those two factors were chosen because they are general factors that are considered before planning for SNF management for a nation.

3.1. Technical considerations

Technical considerations involve many aspects, including disposal method, management method, and SNF transportation safety. This is important due to the fact that improper handling of SNF can cause widespread damage to the people, environment and can incur high costs [4]. In Europe, geological disposal of radioactive wastes is acknowledged by the European Commission (reflected in current EU legislation in 2011) as being widely accepted at the technical level as the most appropriate and sustainable solution [16].

The two methods used to dispose of SNF from research reactors include permanent deep geological disposal [16, 17] and permanent boreholes disposal [8, 16]. Both of these methods are suitable for high-level radioactive wastes such as SNF due to underground burial criteria and existence of a natural engineered barrier system [16]. However, several steps are required to be taken before such permanent disposals can be made, in particular temporary storage in SNF pools, reprocessing of SNF and conditioning treatment necessary for SNF. Not only that, there are other factors, including waste storage design, transport safety and movement plan, which are all important aspects to determine the optimal method for SNF disposal for the research reactor in Malaysia.

3.1.1. Permanent geological facility.

A permanent and safe solution is required in an effort to separate the public and the environment from radioactive wastes produced as a result of energy generation in nuclear reactors, including uses in research, industry and medicine. This step is called a final disposal step, and the steep costs associated with such a step leads to the fact that only a select few countries possess the capabilities to process these wastes. For most countries, the small but significant amount of nuclear waste stemming from research reactor SNF makes the construction of geological repositories non-viable. For countries like these, access to a regional or international repository would be the ideal solution. However, most repositories are designed and licensed to handle the waste from low enrichment uranium, which are the primary fuel for most nuclear power plants [6]. As studies regarding the disposal of high enrichment levels of research reactor fuel in these geological repositories are few and far between, pre-treatment and dilution of research reactor SNF will be required before further studies can clarify the consequences of such actions.

A geological disposal would entail transporting radioactive waste into underground facilities located within a stable geology [16]. This enables medium or high-level radioactive waste to be isolated from the biosphere above ground for the long term [17]. The depth of such facilities will depend mainly on the climate, conditions of underground water, rock formation stability, material composition of the disposal site and threat level of the disposed waste. The safety of the site and experience in handling them will vastly improve public confidence in the radioactive waste management plans.

These deep geological facilities require advanced engineering regardless of its location, and they generally fulfill two functions [17]: The prevention of long-lived radionuclides from migrating near shallow surfaces to put people and environment at risk; and the extensive separation of high-activity waste from the natural surface process and from actions by future generations.

The radioactive wastes and SNF will be arranged underground and closed off when the facilities reach max capacity. However, the permanent closing process for the facility in question will take place over a long period of time, up to hundreds or thousands of years. This time scale necessitates the existence of a regulatory body to evaluate the safety of the site by taking into account natural and anthropogenic processes over a huge time scale [7-9].

3.1.2. Spent nuclear fuel package design.

SNF is carried and wrapped using a specialised package for high-level radioactive wastes [6]. This is of course done to ensure no contamination occurs with its surroundings and the people around it. High quality and effective packages are essential to the safe and sustainable transport, storage and disposal of radioactive wastes. Several requirements are set out by the IAEA and Malaysia has adopted the requirements for such a package in order for it to withstand postulated transportation accidents, such as a 9-m free fall on a hard surface, a drop of 1-m to on a flat pin, withstanding a flame at a temperature of 800°C in 30 minutes and immersion in water up to 15 meters deep [18].

In the same regulations [18], other traits designed to enable safe transportation include long term stability in the presence of radiation, as well as access to its contents to enable gas sampling. There are many listed countries that provide research reactor SNF package for transportation and storage such as Kazakhstan, Latvia, Libya, Poland, Romania, Russian Federation, Serbia, Uzbekistan, Belarus, Bulgaria, Czech Republic, Hungary, Ukraine, Germany, China, Ghana, Viet Nam, Australia, Belgium, Denmark, France, Italy, Portugal, USA, Bolivarian Republic of Venezuela, Netherlands and Sweden [15]. These countries provide the package that can appropriately be used for several modes of transportation such as road, rail, water and air.

3.2. Society and politics

Social factors place their focus on three elements, which are policies and regulations, international and regional cooperation, and public perception, respectively [14, 19]. These factors are crucial in the effort to develop comprehensive options for the management of SNF from research reactors as not only do they refer to technical considerations, but non-economic factors are also taken into account as they can affect the viability of a management plan.

When considering social factors, cost estimation due to risk is difficult to be measured. For example, a lack of legal basis or adequate guidelines can have a significant impact on life cycle costs. On the contrary, rather than trying to assess uncertainty to capture the potential implications of those risks, factors that are likely to impact the implementation of the proposed scenario should be considered in a qualitative way which is also emphasized in this study. The following is some factors that are considered based on guidelines regarding evaluation for innovative nuclear energy systems by the International Project on Innovative Nuclear Reactors and Fuel Cycles (INPRO) [20]: policies and regulations; industrial and technical considerations; political support; public perception; human resource; environmental impact; regional and internal partnerships.

3.2.1. Policies and regulations.

Policies and regulations are guidelines and requirements developed by local regulatory bodies for a spent nuclear fuel disposal plan [5, 7]. These guidelines should be consistent with international rules and treaties while at the same time still tailored to a specific country's need. In Malaysia, Act 304 acts as the premier baseline for the development of policies and regulations regarding SNF from research reactors, in particular the license for borehole disposals and policy statements for national radioactive waste and SNF management [21-23]. In terms of international regulatory commitments, this includes the Treaty on the Non-Proliferation of Nuclear Weapons (NPT) and the Joint Convention on the Safety of Spent Fuel Management and on the Safety of Radioactive Waste Management [24].

In France, the National Plan for Radioactive Materials and Waste Management (PNGMDR) is a key tool in ensuring the long-term implementation of these principles, within the framework set out in the Environment Code and in the Programme Act of 28th June 2006 concerning the sustainable management of radioactive materials and waste [25]. That 3-year plan aims to provide a regular picture of the management policy for radioactive substances nationwide, to assess needs that are coming out and determine objectives to be met in the future, more specifically in terms of studies and research. The interest of this approach was confirmed at European level by the directive establishing a community framework for the responsible and safe management of SNF and radioactive waste, adopted on 19th July 2011, which generalised the adoption of such approach [25].

3.2.2. *Industrial and technical considerations.*

Considerations must be given to technological and industrial capacity when making choices on radioactive waste management plans [5, 7, 9, 13]. Options that require more extensive investment in new facilities and technological development can be seen negatively because it involves risks such as delays and unexpected issues, but it may also be viewed positively as it represents an opportunity for investment in national infrastructure and technological capacity [13]. A few considerations may include the following: presence of suitable machinery for projects; availability of basic and technical infrastructure such as roads, bridge and emergency services; whether the option provides value or places a burden on the national industry; suitability of locations for storage facilities [12].

3.2.3. *Political support.*

Any organization responsible for SNF disposal should be in constant contact with the government regarding decisions on choices and strategy [19]. This is to enable the government to underline the needs of the nation and to determine more practical methods for the aforementioned strategies [5, 6, 7, 8. 12, 13, 15]. In the end, the government will be the party who has the last say for decisions while taking into account opinions from stakeholders.

3.2.4. *Public perception.*

Closely related to political support is the fact that opinions and perceptions of the general public plays a pivotal role in the selection of SNF management [12, 13, 19]. Hence, the public's perception of the many options available has to be properly evaluated and understood to enable a governmental commitment to support such a choice. France also underwent an environmental assessment and a public consultation to reinforce the consideration given to environmental aspects, while reminding the public the importance and beneficial purpose of their national plan to be responsible and safe management of SNF and radioactive waste [25].

Generally, the public would be concerned about both the environment and them, and the perception that there are no safe methods to store SNF and radioactive wastes owing to the risks associated with the wastes. While it is difficult to convince the public of the safety of the proposed management plans, there are several factors that influence the perception and risk acceptance by individuals, such as trust in institutions, individual risk assessment, degree of control, reward to risk considerations, knowledge and experience, and potential for natural disasters [14, 19]. The government can alleviate such perceptions through effective transmission of information, being attentive to the opinions of the public, decision with consultation, and a show of commitment from the government [12, 19].

This is the most challenging issue that needs to be addressed for a nation in deciding on SNF management. Public confidence with regard to the management of SNF is heavily influenced by perceived risks as well as a lack of confidence in, or understanding of, SNF management [14]. The public is still haunted by the effects from the Chernobyl disaster and following the Fukushima accident that released radiation which they think "it can happen again".

3.2.5. *Human resource.*

A robust pool of human resources, both in terms of knowledge and quantity are needed in order to achieve the desired options. Qualified staff in sufficient numbers are essential to the successful implementation of the SNF disposal plans as they would be able to fulfil the diverse needs of the various facilities, whether governmental or regulatory [19]. Amongst the various fields that require the presence of high-quality human resources include nuclear waste facility operations, their maintenance and their decommissioning [12, 14]. The lack in human resources can be rectified through education programs, or through imports to replace older expectations according to the needs of the state [4, 6].

3.2.6. Environmental impact.

All scenarios pertaining to the management of SNF must be evaluated for compliance towards environmental regulations. The costs associated with the necessary steps including environmental preservation and recovery must be included in the cost estimates [6, 12]. Some factors to be considered include the relevant environmental regulations for each choice made, the responsible agency for environmental law enforcement, experts on the impact of national environmental policy, perception of the public regarding its impact on the environment, and non-governmental organisations (NGOs) that could influence or block certain decisions.

3.2.7. Regional and international partnerships.

Existing or planned regional and international partnerships can directly influence the required domestic source and infrastructure for each individual choice [12]. For example, when the laws and regulations between two nations are synchronized regarding nuclear and radiological activities, the friction when making choices between them are significantly reduced [14]. This can provide substantial benefits for the supplier and user of technologies as these treaties create convention, standards and international guidelines. These kinds of cooperation will also reduce the amount of infrastructure needed by adopting similar institutions, industrial infrastructure, and appropriate international norms and standards [12].

4. Conclusion

This study systematically examined the management method of spent nuclear fuel from research reactors and identified the factors that influence the disposal plan in Malaysia. The results of various analyses show that two factors influence plans for the disposal of spent nuclear fuel from research reactors in Malaysia. Technical factors, mainly the proposed nuclear fuel disposal method applicable in Malaysia. This study has proposed two methods of fuel disposal used nuclear from research reactors in Malaysia which is the deep geological disposal method and the deep borehole disposal method, in addition to other requirements such as waste barrels. The second factor is the society and politics factor, specifically non-economic factors such as policies and regulations, industrial and technical considerations, political support, public perception, human resource, environmental impact, as well as regional and international partnerships. Further examination of other factors such as economic and site selection can further contribute to the development of the SNF disposal plans in Malaysia.

5. References

[1] United Nations. 2024. Treaty on the non-proliferation of nuclear weapons: Office for disarmament affairs treaties database. https://treaties.unoda.org/t/npt/participants?group=member-states [5 October 2024]
[2] Nuklear Malaysia 2024 TRIGA PUSPATI Reactor. https://www.nuclearmalaysia.gov.my/eng/kemudahan-rnd.php?id=1 [24 August 2024]
[3] Gefen, D. & Straub, D. 2005. A Practical Guide to Factorial Validity Using PLS-Graph: Tutorial and Annotated Example. Communications of the Association for Information Systems: 16: 91-109.
[4] IAEA. 2014. General Safety Requirements GSR Part 3, Radiation Protection and Safety of Radiation Sources: International Basic Safety Standards. Vienna: International Atomic Energy Agency.
[5] IAEA. 2009. General Safety Requirements GSR Part 5, Predisposal Management of Radioactive Waste. Vienna: International Atomic Energy Agency.
[6] IAEA. 2017. Specific Safety Requirement SSR-4, Safety of Nuclear Fuel Cycle Facilities. Vienna: International Atomic Energy Agency.
[7] IAEA. 2011. Specific Safety Requirement SSR-5, Disposal of Radioactive Waste. Vienna: International Atomic Energy Agency.

[8] IAEA. 2024. Specific Safety Guide SSG-1 (Rev. 1), Borehole Disposal Facilities for Disused Sealed Radioactive Sources. Vienna: International Atomic Energy Agency.
[9] IAEA. 2011. Specific Safety Guide SSG-14, Geological Disposal Facilities for Radioactive Waste. Vienna: International Atomic Energy Agency.
[10] IAEA. 2020. Storage of spent nuclear fuel. Specific safety guide No. SSG-15 (Rev. 1). Vienna: International Atomic Energy Agency (IAEA).
[11] IAEA. 2017. Specific Safety Guide SSG-42, Safety of Nuclear Fuel Reprocessing Facilities. Vienna: International Atomic Energy Agency.
[12] IAEA. 2009. Policies and Strategy for Radioactive Waste Management, Nuclear energy Series No. NW-G-1.1. Vienna: International Atomic Energy Agency.
[13] IAEA. 2018. Options for management of spent nuclear fuel and radioactive waste for countries developing new nuclear power programmes. Nuclear Energy Series No. NW-T-1.24. Vienna: International Atomic Energy Agency.
[14] IAEA. 2019. Storing Spent Fuel until Transport to Reprocessing or Disposal, Nuclear Energy Series No. NF-T-3.3. Vienna: International Atomic Energy Agency.
[15] IAEA. 2017. Available Reprocessing and Recycling Services for Research Reactor Spent Nuclear Fuel, Nuclear Energy Series No. NW-T-1.11. Vienna: International Atomic Energy Agency.
[16] Chapman, N. & Hooper, A. 2012. The disposal of radioactive waste underground. Proceedings of the Geologists' Association 123: 46-63.
[17] Kvashnina, K., Claret, F., Clavier, N., Levitskaia, T. G., Wainwright, H. & Yao, T. 2024. Long-term, sustainable solutions to radioactive waste management. *Scientific Reports* 14: 5907.
[18] Malaysia. 1989. Atomic Energy Licensing Act (Act 304), Radiation Protection (Transport) Regulations.
[19] IAEA. 2007. Factors Affecting Public and Political Acceptance for the Implementation of Geological Disposal, IAEA-TECDOC-1566. Vienna: International Atomic Energy Agency.
[20] IAEA. 2003. Guidance for the evaluation of innovative nuclear reactors and fuel cycles Report of Phase 1A of the International Project on Innovative Nuclear Reactors and Fuel Cycles (INPRO), IAEA-TECDOC-1362. Vienna: International Atomic Energy Agency.
[21] Malaysia. 1984. Atomic Energy Licensing Act (Act 304).
[22] Malaysia. 2011. Atomic Energy Licensing Act (Act 304), Radiation Protection (Radioactive Waste Management) Regulations.
[23] Malaysia. 2021. Pernyataan Dasar Pengurusan Sisa Radioaktif dan Bahan Api Nuklear Terpakai Kebangsaan LEM/DASAR/3.
[24] IAEA. 1997. Joint Convention on the Safety of Spent Fuel Management and on the Safety of Radioactive Waste Management, INFCIRC/546. Vienna: International Atomic Energy Agency.
[25] PNGMDR. 2017. French National Plan for the management of Radioactive Materials and Waste 2016-2018. Ministry for an Ecological and Solidary Transition.

IOP Conf. Series: Materials Science and Engineering **1326** (2025) 012008 doi:10.1088/1757-899X/1326/1/012008

Extraction of rare earth element from local (Malaysian) monazite mineral

C Rohaida Che Hak[1,a], Nurliana Roslan[1,b]

[1]Material Technology Group, Industrial Technology Division, Malaysian Nuclear Agency, 43000 Kajang Selangor.

[a]rohaida@nm.gov.my, [b]nurliana_r@nm.gov.my

Abstract. Rare earth elements (REEs) are 15 elements from lanthanide group and 2 elements from transition metal group in periodic table. REEs have been identified as advanced materials because of their peculiar properties. It can be used in wide range of applications especially in electronic devices, electric cars and chemical processing catalysts. In Malaysia, REEs are mainly found in monazite, a kind of phosphate mineral. It consists of more than 60 wt.% total concentration of cerium, lanthanum, praseodymium, and neodymium. Monazite also contains silicon, iron and thorium and uranium. Because of the complexity of this mineral, extraction of REEs requires various steps of hydrometallurgical processes and each step needs several controlled parameters. This paper explained step-by-step processes of cerium separation and extraction from monazite. This study used alkaline cracking technique to digest monazite and remove soluble phosphates to recover REEs in solid hydroxide cake. Cerium was separated from other REEs via oxidation and selective precipitation. The samples were analysed using Energy Dispersive X-Ray Fluorescence (EDXRF) and Field Emission Scanning Electron Microscope - Energy Dispersive X-Ray spectrometer (FESEM-EDXs) to determine the presence of REEs at every step of processing. Results showed that cerium can be extracted from monazite mineral via those defined processes.

1. Introduction

REEs consist of 17 elements, 15 are from lanthanide group with atomic number from 57 to 71 and other 2 are from transition metal group with atomic number of 21 and 39. REEs generally are divided into two groups that is defined by "light REEs" (LREEs) with atomic number from 57 to 64 which is lanthanum (La), cerium (Ce), praseodymium (Pr), neodymium (Nd), promethium (Pm), samarium (Sm), europium (Eu) and gadolinium (Gd). The other group is defined as "heavy REEs" (HREEs) with atomic number of 65 to 71 which is terbium (Tb), dysprosium (Dy), holmium (Ho), erbium (Er) thulium (Tu), Ytterbium (Yb) and lutetium (Lu). Scandium (Sc) and Yttrium (Y) with atomic number of 21 and 39 respectively is classified as HREEs because their similarities in chemical properties [1,2]. REEs have been identified as advanced materials because of their special properties which enable the REEs to be used in wide range of applications such as in electronic devices for smart phone and tablets, manufacturing of various components in electric and hybrid cars, autocatalytic converters for advanced chemical processing, medical imaging, wind turbine for renewable and nuclear energy, additives for in alloy and metallurgy and so forth [3,4].

In Malaysia, the primary source for REEs are monazite and xenotime which are obtained from tin mining by-products (*amang*). These rare earth minerals are concentrated by beneficiary processes such as flotation, magnetic or gravity methods to produce rare earth concentrates [5]. In the previous studies, Malaysian monazite was found to contain mainly LREE with chemical formula of [Ce, Nd, La]-PO_4 and the content of REEs concentrates is dependent on the particle size. For monazite with an average size of 6 μm has higher total concentration of REEs which is 60 wt%, while monazite with average size of 180 μm has 50 wt.% of REEs. For smaller size monazite the REEs present are Ce, La, Nd, Gd, Sm, Pr, Dy Yb, Er and Y, while for larger size the REEs present are Ce, La, Nd, Gd and Y. Morphology analysis shows that the REEs have higher concentration in dense particle. This suggest that it is important to pre-treat monazite particles to reduce their size and the density, thereby exposing more REEs in the mineral and enhancing the digestion process [6]. Since the complexity of the mineral, REEs must be extracted in a certain way to ensure that just the desired element is obtained. The process involves with alkaline digestion, acid leaching and further extraction process to purify the REEs.

Alkaline digestion is a preferred method for processing of monazite because stems of its ability to remove phosphates as well as the simplification of the thorium concentrate purification phase [7], does not produce any liquid wastes, low maintenance cost, and able to regenerate the alkali. It involves treating monazite with sodium hydroxide (NaOH) at elevated temperatures, typically between 140-150 °C. The process breaks down the mineral structure and then converts rare earth phosphates into rare earth hydroxides, which are more easily separated and purified. The general reaction can be represented as Equation 1:

$$REPO_4 + 3NaOH \rightarrow RE(OH)_3 + Na_3PO_4 \quad (1)$$

where RE represents rare earth elements.

After digestion, a water leaching or washing process is necessary to remove soluble phosphates and excess NaOH, while acid leaching is required to dissolve the REEs from hydroxide cake. Acid leaching with HCl process is effective and simple to generate REEs in liquor solution. The best leaching conditions are typically achieved at 60-80°C with 3-6 M HCl, resulting in high recovery rates, often over 90% for most REEs [7,9,11,12 ,13,14]. The general reaction for the acid leaching step can be represented as Equation 2:

$$RE(OH)_3 + 3HCl \rightarrow RECl_3 + 3H_2O \quad (2)$$

After acid leaching, Ce is the first element to be separated and extracted from the REEs in liquor solution. Ce is one of the most abundant REEs found in monazite, typically constituting about 45-50% of the total REE content in Malaysian monazite as reported in the previous study [6] and the content is dependent on the particle size as well [10]. However, the drawback for alkaline cracking is that the particles size of monazite must be below 325 mesh (44μm) to achieve a high yield of hydroxide REEs. Smaller monazite particles have a larger surface area-to-volume ratio, which increases the reactivity with alkaline solution and leads to more efficient phosphate conversion [9]. In addition to that, Ce is the only is the only lanthanide stable in the +4 oxidation state in aqueous solutions due to its lowest standard electrode potential of Ce(IV)/Ce(III) [15] and has a strong coordination ability with anions such as F^-, NO_3^-, HSO_4^-, and $H_2PO_4^-$ in solution because of its small ionic radius, making it easy to be extracted by organic extractants [16]. Therefore, in this study a pre-treatment of monazite was carried out using industrial roll ball milling prior to digestion process to get the smallest possible of monazite particles. Then for Ce separation from other REES, it is done by selective oxidation to convert Ce to Ce (IV) by

IOP Conf. Series: Materials Science and Engineering **1326** (2025) 012008 doi:10.1088/1757-899X/1326/1/012008

chemical oxidation using permanganate. The separation of Ce through the oxidation process is typically followed by the selective dissolution of the trivalent REEs to keep the cerium (IV) hydroxide in its insoluble form.

Therefore, the objective of this study is to develop techniques for the separation and extraction of Ce from Malaysian monazite, beginning with the pre-treatment of the minerals, followed by alkaline digestion to remove phosphate, acid leaching to generate REE-enriched leach solution and selective oxidation and precipitation to separate Ce from other REEs.

2. Methodology

The high-grade monazite (95%) was obtained from a factory in Perak, northern area of Malaysia. All chemicals and reagents used were analytical grade and were prepared using distilled water. The methodologies that developed in this work is simplified as flow chart shown in Figure 1.

2.1. Pre-treatment

Monazite was ground (pre-treated) using industrial roll ball milling for about 5 hours to obtain fine particles mineral. The method of grinding and particle size measurement was previously reported in Roslan et al. [10].

2.2. Digestion and washing

A sodium hydroxide (NaOH) solution was prepared by mixing NaOH with distilled water. Monazite was weighted and mixed with NaOH solution at temperature of about 100–150 °C, and the mixture was stirred for 30 minutes to 1 hour. Following digestion, the slurry was diluted with distilled water, heated and agitated for 1 hour at 100°C before being allowed to cool down. During digestion, phosphorus is converted into soluble phosphate, while the remaining solid consists of a hydrous rare earth cake that is still in monazite phase. To ensure complete conversion of phosphorus into preferred trisodium phosphate (TSP) and to remove all free NaOH, the washing steps must be repeated several times until a neutral pH is achieved.

2.3. Phosphate removal

The leached solution was filtered to separate the insoluble hydroxide cake from the soluble phosphate. The hydroxide cake was dried in the oven at 70°C and collected to measure its yield. The filtered solution, which contained soluble phosphate was evaporated until the volume reached one-third of the total amount. The evaporated solution was then filtered, and the resulting white crystalline powder was dried in an oven for further analysis.

2.4. Acid leaching

The dried hydroxide cake was weighed and mixed with hydrochloric acid (HCl). The prepared solution was heated and stirred for 1 hour, with the temperature kept below 70°C and the pH was maintained between 2.9 and 3.1. The leaching product was filtered to separate the rare earth chloride ($RECl_3$) solution from the second generation of hydroxide cake. The solid was then dried in an oven, while the filtrate was kept for further processing.

2.5. Oxidation

Potassium permanganate ($KMnO_4$) and sodium carbonate (Na_2CO_3) solution were prepared separately by dissolving each-compound in distilled water. A certain amount of $RECl_3$ at pH 6.06 was mixed with $KMnO_4$ and Na_2CO_3 under three different conditions:

(i) $KMnO_4$ was added until the pH remained in the range of 2.5-3.5. Stirring continued for 1 hour until the final pH recorded was 3.17 and then the mixture was allowed to precipitate. The precipitate was collected through filtration and dried in an oven for further analysis, leaving other REEs in solution (supernatant).

(ii) The supernatant from (i) was treated with $KMnO_4$ followed by stirring for 1 hour. The recorded pH was 4.9. Na_2CO_3 was then added at certain amount and the final pH was recorded at 5.40. The resulting precipitate was collected and dried in an oven for further analysis.

(iii) The supernatant obtained from filtration in (ii) was treated with an excess of $KMnO_4$ followed by stirring for 1 hour. Na_2CO_3 was added and the final pH recorded was 4.62. The precipitate was collected and dried in an oven for further analysis. The remaining supernatant was kept for further analysis.

2.6. Characterization

The particle size of pre-treated monazite processed using ball milling was measured using particle size analyser (Microtrac-X100). Samples from each processing steps were collected and analysed using various techniques including microscopic analysis using FESEM (Carl Zeiss, GeminiSEM 500), elemental analysis using energy EDXs (Oxford Instruments, X-MAX 80) and elemental analysis of all filtrate solutions and supernatants using EDXRF (Shimadzu). Some solid samples were also analysed for phase analysis using X-Ray diffractometer (XRD, Panalytical, X-Pert Pro).

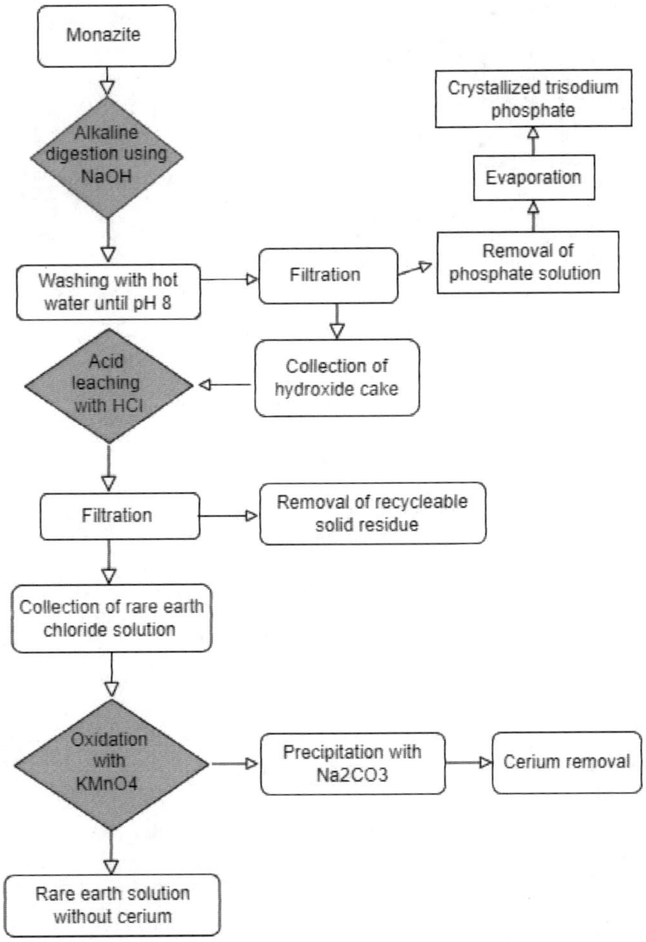

Figure 1. Flowchart of monazite processing to remove cerium from other rare earth elements

IOP Conf. Series: Materials Science and Engineering **1326** (2025) 012008 doi:10.1088/1757-899X/1326/1/012008

3. Results and discussion

3.1. Pre-treatment of monazite

Figure 2 illustrates the physical properties of pre-treated monazite, which shows a non-homogenous distribution of particle size after milling in an industrial roll mill for 5 hours (Figure 2a). This is supported by the particle size distribution analysis shown in Figure 2b, which indicates that 89% of the monazite particles have a diameter of 5.12 μm, while 11% have a diameter of 0.685 μm. The initial particle size of the monazite was measured at 194.5 μm, indicating that the milling process reduced the particle size by approximately 90%. In the previous studies, the effect of the pre-treatment method to the REEs recovery from monazite samples was investigated using milling and sieving process. Results showed that the treated monazite particles had smaller size with inhomogeneous distribution, higher composition of the major REE phosphates (Ce, La, and Nd) compared to the non-treated samples. The composition (wt.%) of these elements in the monazite mineral was confirmed using XRD analysis [10].

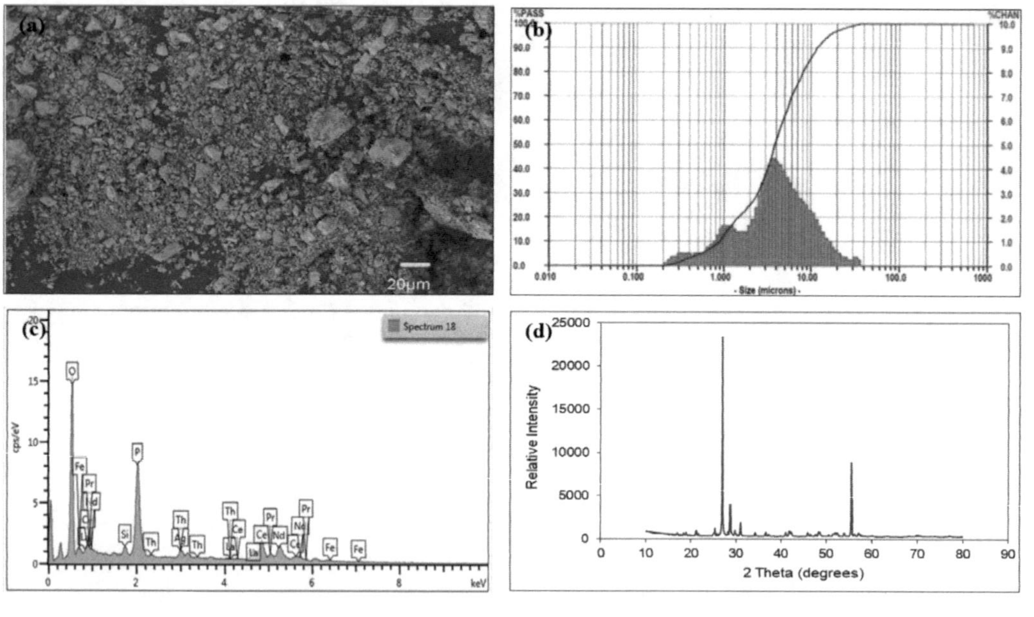

Figure 2. (a) Surface morphology, (b) particle size distribution, (c) EDX spectrum and (d) XRD spectrum of milled monazite.

The EDX spectrum of monazite shown in Figure 2c confirms that monazite is a rare earth phosphate mineral, as indicated by elemental composition presented in Table 1. The elemental composition of monazite shows Ce is the major element, followed by La, Nd and Pr. The total REEs concentration which is the sum concentration of all REEs present in monazite is 53.56 wt.% for Ce, La, Nd and Pr, the rest are another metal element such as P, Si, Fe and Ag as shown in Table 1. Thus, XRD phase analysis confirms that the monazite particles are rare earth phosphates, identified as monazite phase (Figure 2d). These results show that Malaysian monazite contains LREEs, with Ce being the majority and Pr is the minority and contain about 6 wt.% of Th. In the previous work, it was reported that Th can be separated from REEs and extracted via solvent extraction technique up to 85 wt.%. However, further purification to 99 wt.% is required for Th to be suitable for use in nuclear energy reactor [20].

IOP Conf. Series: Materials Science and Engineering **1326** (2025) 012008 doi:10.1088/1757-899X/1326/1/012008

3.2. Monazite digestion and washing

Figure 3a shows the yellowish-brown hydroxide cake obtained after monazite was digested with NaOH. EDX analysis indicates that 64.4 wt.% of total REEs (Ce, La, Nd, Pr and Sm) was detected in hydroxide cake, compared to 53.56 wt.% detected in monazite before digestion as shown in Table 1. Before digestion, only 4 types of REEs were observed in monazite which are Ce, La, Nd and Pr, while Sm was also detected after digestion. The increase in REE content is attributed to the dissolution of other elements, such as phosphorus (P), iron (Fe), silver (Ag), and silicon (Si), during digestion, as none of these elements were detected in the EDX analysis of the hydroxide cake (Table 1). The dissolution of P, Fe, Ag and Si increased the rare earth content in hydroxide cake, while the amount of Th also increased. In the discussion, this hydroxide cake is referred to as rare earth hydroxide ($REOH_3$). For Ce extraction from monazite, the alkaline process is often preferred due to its effectiveness in separating Th from REEs [8].

Table 1. Elemental composition of monazite, hydroxide cake after digestion, residue after acid leaching and TSP measured using FESEM-EDX

Element	Milled monazite ($RE(PO)_4$)	Hydroxide cake after digestion ($RE(OH)_3$)	Residue cake after acid leaching	Trisodium phosphate (Na_3PO_4)
	Concentration (wt.%)	Concentration (wt.%)	Concentration (wt.%)	Concentration (wt.%)
Cerium (Ce)	27.48 (±0.66)	31.90 (±0.40)	21.40 (±0.28)	-
Lanthanum (La)	11.97 (±0.52)	13.20 (±0.40)	9.05 (±0.24)	-
Neodymium (Nd)	11.01 (±0.62)	12.90 (±0.50)	7.70 (±0.30)	-
Praseodymium (Pr)	3.10 (±0.59)	3.50 (±0.40)	-	-
Samarium (Sm)	-	2.90 (±0.50)	-	-
Oxygen (O)	27.78 (±0.52)	27.78 (±0.52)	31.45 (±0.22)	39.9 (±0.20)
Carbon (C)	-	3.0 (±0.10)	6.56 (±0.19)	12.6 (±0.40)
Phosphorus (P)	10.15 (±0.22)	-	7.59 (±0.58)	0.8 (±0.10)
Iron (Fe)	1.43 (±0.28)	-	-	-
Silicon (Si)	1.09 (±0.07)	-	0.55 (±0.30)	10.9 (±0.10)
Silver (Ag)	0.21 (±0.67)	-	-	-
Calcium (Ca)	-	0.70 (±0.10)	0.71 (±0.05)	-
Chlorine (Cl)	-	-	1.49 (±0.04)	9.8 (±0.10)
Sodium (Na)	-	-	-	21.6, (±0.10)
Potassium (K)	-	-	-	2.5 (±0.10)
Sulphur (S)	-	-	-	1.8 (±0.10)
Thorium (Th)	5.77 (±0.95)	7.60 (±0.20)	13.50 (±0.21)	-
Total REE	53.56	64.41	38.15	0

Figure 3b shows the white powder obtained after evaporation of phosphate solution. The elemental composition of this white powder as shown in Table 1, indicates the presence of sodium (Na), silicon (Si), chlorine (Cl), sulphur (S), potassium (K) and phosphorus (P). None of REEs and Th were observed, suggesting that all REEs, and Th were effectively separated from the phosphate solution during digestion. This result confirms that P was dissolved during digestion, along with other metals such as Fe, Ag, and Si, as well as other impurities, which were no longer detected in $REOH_3$. Figure 3g presents the XRD spectrum of the powder, indicating that the phase analysis corresponds to 86.3 wt.% of trisodium phosphate (TSP). This is sufficient to conclude that the monazite alkaline digestion successfully produced TSP powder. The electron images shown in Figures 3d and 3e illustrate the morphological differences between the two phases, $REOH_3$ and TSP.

3.3. Acid dissolution

The reaction between dried hydroxide cake and 37% HCl dissolved the REEs into a chloride solution, while the undissolved elements remained as residue after filtration. Figure 3c shows the pale-brown residue cake obtained after acid leaching. EDX analysis of this powder indicates that the total REE content decreased to 38.15 wt.%, a lower amount compared to the total REE content in monazite and $REOH_3$, as shown in Table 1. The REEs present in this residue are Ce, La, and Nd, while the Th content increased to 13.5 wt.%, which is higher than in monazite and $REOH_3$. The morphology of this residue is more porous compared to $REOH_3$, likely due to the acid leaching process, which caused the dense particles to dissolve. The remaining REEs and Th suggest that this residue is still suitable for further acid leaching processes to recover the valuable elements.

Figure 3. (a) hydroxide cake powder obtained after digestion ($REOH_3$), (b) white powder obtained after evaporation of phosphate solution (TSP), (c) solid residue after acid leaching, (d) morphology of $REOH_3$, (e) morphology of TSP, (f) morphology of solid residue, (g) XRD spectrum of TSP.

Analysis of the filtrate using EDXRF shows that approximately 94.57 wt.% of rare earth ions were dissolved in the solution after acid dissolution, as shown in Table 2. The main elements present are Nd, followed by Ce, La, Pr, and Y. This solution is referred to as rare earth chloride solution ($RECl_3$) or REE pregnant leach solution, which will be used as the mother liquor for the subsequent REE separation steps. No Th was detected in the $RECl_3$ solution, indicating that the radioactive element was separated into the undissolved solid. These results demonstrate that the acid leaching successfully extracted nearly 95% of the LREEs from the hydroxide cake, while Th was separated from the REE liquor and retained in the solid residue.

Table 2. Elemental composition of $RECl_3$ measured and final solution after oxidation measured using EDX fluorescence

Element	$RECl_3$	Final solution after filtration of third oxidation
	Concentration (wt.%)	Concentration (wt.%)
Neodymium (Nd)	36.39	-
Cerium (Ce)	34.51	-
Lanthanum (La)	14.20	-
Praseodymium (Pr)	6.30	-
Yttrium (Y)	3.17	-
Phosphorus (P)	2.65	0.004
Calcium (Ca)	2.15	0.005
Thulium (Tm)	0.63	-
Manganese (Mn)	-	0.059
Potassium (K)	-	0.008
Copper (Cu)	-	0.001
Oxygen (O)	-	-
Total REE	94.57	0

3.4. Cerium extraction via oxidation

A certain amount of $RECl_3$ was mixed with $KMnO_4$ to oxidize Ce from its trivalent form to its tetravalent form. The oxidation of Ce utilizing $KMnO_4$ caused the $RECl_3$ solution to turn purple, and after being left overnight, a precipitate formed at the bottom of the colorless solution, as shown in Figure 4. At this step, three samples were collected at different pH: pH 3.17, pH 4.62 and pH 5.4. For pH 3.17, no Na_2CO_3 was added, for pH 4.62 both $KMnO_4$ and Na_2CO_3 were added but the amount of $KMnO_4$ was excessive, while for pH 5.4 both $KMnO_4$ and Na_2CO_3 were added at reasonable amount. The colour of precipitates varied depending on the pH; yellowish for pH 3.17, purplish at pH 4.62 and pink brownish at pH 5.4.

The morphology and EDX spectrum of the precipitate at pH 5.4 are shown in Figure 4e and 4f respectively, while the elemental composition analysis is presented in Table 3. The highest amount of Ce separated was at pH 5.4, with a concentration of 57.46 wt.%, followed by 55.98 wt.% at pH 3.17. A significant amount of manganese (Mn), Na and Cl was detected at both pH levels, indicating that thorough washing is necessary to remove the excess of these elements. Th was observed in the sample at pH 3.17 but not in the sample at pH 5.4, suggesting that pH 3.17 or lower favors the precipitation of more Th. At pH 4.62, Ce was not detected, but Nd, La, Sm, and Pr were present in the dried solid, with a total REE content of 20.92 wt.%. The reason of no Ce was detected in the third oxidation is possibly because most Ce was extracted in the first oxidation (at pH 3.17) and in the second oxidation (at pH 5.4). After the solid was collected after third oxidation, the final supernatant was analysed with EDXRF, results indicated that no rare earth ions were detected in the solution as shown in Table 2. The major element in the final solution is Mn. Therefore, these results can be inferred that oxidation of $RECl_3$ with $KMnO_4$ was successfully separated the Ce compound from the solution in the first and second stages oxidation. Further addition of $KMnO_4$ and Na_2CO_3 into the supernatant did not increase Ce concentration but Nd, La, Sm and Pr was precipitated.

This observation aligns with findings reported by Abreu & Morais [17], where the conversion of Ce from Ce^{3+} to Ce^{4+} was accomplished by adding a mixture of $KMnO_4$ and Na_2CO_3 to the RECl solution. The precipitation process resulted in the recovery of solid ceric hydroxide and manganese dioxide. In their study, during oxidation, the pH was adjusted to around 3-4 to optimize the oxidation process. The oxidation process is highly selective for Ce due to its unique redox behavior where the standard reduction potential for Ce (IV)/Ce (III) is about +1.72 V, significantly different from other lanthanides, enabling its selective oxidation. The oxidation of Ce (III) to Ce (IV) occurs relatively fast compared to other REEs, contributing to its selectivity. In their study, the cerium hydroxide was treated with oxalic acid to selectively precipitate 98.1% of Ce oxalate (CeO_2) and only 0.07 wt.% manganese [17].

The efficiency of separation depends on factors like pH, temperature, and oxidant concentration. In some processes, complexing agents are added to stabilize Ce (IV) in solution, allowing for subsequent separation techniques like solvent extraction [9,18,19]. According to Soe et.al [21], selective precipitation with ammonium hydroxide is required to precipitate rare earth hydroxide from $RECl_3$ at pH 11. Following this selective precipitation, the rare earth hydroxide needs to be digested with nitric acid before being precipitated with $KMnO_4$ and Na_2CO_3 to form cerium carbonate at pH 3.8.

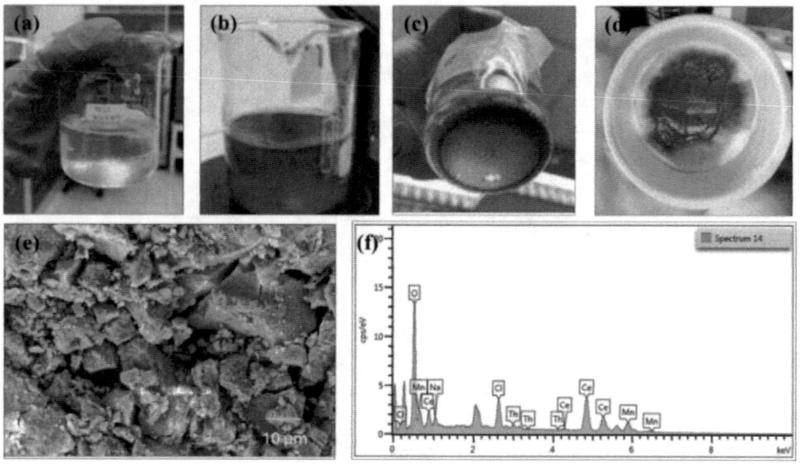

Figure 4. (a) $RECl_3$ solution, (b) $RECl_3$ solution after addition of $KMnO_4$, (c) Precipitate at the bottom of beaker, (d) the collected precipitates, (e) morphology of precipitate at pH 5.4, and (f) EDX spectrum of precipitate at pH 5.4.

Table 3. Elemental composition rare earth compounds after oxidation process at different pH measured using FESEM-EDX analysis.

Element	pH 3.17 Concentration (wt%	pH 4.62 Concentration (wt%)	pH 5.4 Concentration (wt%)
Cerium (Ce)	55.98 (±0.42)	-	57.46 (±0.40)
Manganese (Mn)	10.54 (±0.31)	34.29 (±0.56)	9.94 (±0.34)
Thorium (Th)	3.20 (±0.31)	-	-
Oxygen (O)	23.26 (±0.25)	42.59 (±0.59)	26.30 (±0.28)
Chlorine (Cl)	5.64 (±0.11)	2.19 (±0.10)	4.57 (±0.11)
Sodium (Na)	1.37 (±0.10)		1.73 (±0.11)
Neodymium (Nd)	-	11.22 (±0.61s)	-
Lanthanum (La)	-	4.26 (±0.42)	-
Praseodymium (Pr)	-	3.12 (±0.54)	-
Samarium (Sm)	-	2.32 (±0.69)	-
Total REE	55.98	20.92	57.46

4. Conclusion

Ce was separated from Malaysian monazite via alkaline digestion, acid leaching and oxidation processes as shown in Figure 1. The monazite used is a rare earth phosphate mineral consisting mainly of LREE. The monazite particle size used about 5μm and was digested with NaOH. Ce is the major elements in monazite and the first to be separated. The products of digestion were a brown insoluble hydroxide cake and a phosphate solution. The hydroxide cake obtained underwent a leaching process using HCl acid to extract rare earth ions, resulting in a chloride solution enriched with approximately 95% of LREEs. Whereas the phosphate solution required-intensive washing until a pH 8 was reached to remove excess of sodium, followed by evaporation to obtain crystallized white TSP powder as the first product from monazite digestion. Oxidation of the rare earth chloride using $KMnO_4$, followed by precipitation with Na_2CO_3 at first and second oxidation at pH 3.17 and 5.4 respectively, completely removed Ce from the solution, leaving behind Nd, La and Pr. The oxidation pH had to be above 3.17 to avoid Th precipitation. Further work is needed to extract Cerium compounds in the form of carbonate or oxide, which can be utilized in various industries.

5. References

[1] Asadollahzadeh M, Torkaman R, and Torab-Mostaedi M 2020 Separation and Purification Reviews 1

[2] Salfate G and Sánchez J 2022 Polymers 14 21

[3] Balaram V 2019 Geoscience Frontiers 10 (4) 1285.

[4] Anastopoulos I, Bhatnagar A and Lima EC 2016 *J. Mol. Liq.* 221 954

[5] Haldar SK 2017 Platinum-Nickel-Chromium Deposits, 267. Elsevier.

[6] Cik Rohaida CH, Sarimah M, Ismail M and Susan MS 2020 Jurnal Sains Nuklear Malaysia 32(1) 10

[7] Sadri F, Nazari AM and Ghahreman A 2017 J. Rare Earths 35 (8) 739

[8] Kim W, Bae, I, Chae S and Shin H 2009 J. Alloys. Compd. 486 (1-2) 610

[9] Kumari A, Panda R, Jha MK, Kumar JR and Lee JY 2015 A review Minerals Engineering 79 102

[10] Nurliana R, Cik Rohaida CH, Abdul Rahman N 2023 IOP Conf. Series: Materials Science and Engineering 1285 012007

[11] Xie F, Zhang TA, Dreisinger D and Doyle F 2014 Minerals Engineering 56 10

[12] Jha MK, Kumari A, Panda R, Kumar JR, Yoo K and Lee JY 2016 Hydrometallurgy 165 2

[13] Kim E and Osseo-Asare K 2012 Hydrometallurgy 113 67

[14] Zhang J and Edwards C 2013 In 44th Annual Meeting of the Canadian Mineral Processors 79

[15] Zou D, Chen J and Li D 2014 J. Rare Earths 32 (8) 681

[16] Li K, Chen J and Zou D 2019 Cerium Oxide - Applications and Attributes IntechOpen.

[17] Abreu RD and Morais CA 2010 Minerals Eng. 23(6) 536.

[18] Binnemans K, Jones PT, Blanpain B, Van Gerven, T, Yang Y and Walton A 2013 J Clean. Prod. 51 1

[19] Zhang W, Rezaee M, Bhagavatula A, Li Y, Groppo J and Honaker R 2015 Int. J. Coal Prep. Uti. 35(6) 295

[20] Cik Rohaida CH, Zakiuddin J, Ismail M and Sarimah M 2022 Key Engineering Materials 908 509

[21] Soe NN, Shwe LT and Lwin KT, 2008 Int. J. Materials and Metallurgical Engineering 2 10

Acknowledgement

I would like to acknowledge three university students who were completed their internship under my supervision. During the internship, they helped me to perform the experimental work and data collection for this article. The students were M Azizi Tamizi from UiTM Shah Alam, N Adlina Johari from UiTM Kuala Pilah, and Aisyah Jasmin M Al Saari from UPM Serdang. They did their job well and collected many data to make this article a success.

IOP Conf. Series: Materials Science and Engineering **1326** (2025) 012009 doi:10.1088/1757-899X/1326/1/012009

Comparison of the potential risk associated with various SMR designs

Nur Fatin Hanani Razali[1,2] and Norsyahidah Mohd Hidzir[1,2]*

[1]Department of Applied Physics, Faculty of Science and Technology, Universiti Kebangsaan Malaysia (UKM), 43600 Bangi, Selangor, Malaysia.

[2]Nuclear Technology Research Centre, Faculty of Science and Technology, Universiti Kebangsaan Malaysia (UKM), 43600 Bangi, Selangor, Malaysia.

*syahidah@ukm.edu.my

Abstract. Each small modular reactor (SMR) design presents a distinct array of risks due to its specialized technology and systems. The probability of safety assessment (PSA) Level 1 is a technique used to evaluate the safety risks of nuclear power plants, focusing on the probability of reactor core damage. Current research mainly emphasizes the advantages and safety features of a particular SMR design. However, there is a deficiency in the available information and data needed for a thorough comparative analysis of the potential risks associated with various SMR designs, particularly in terms of the identification systems and design technologies that are most critical to safety and present the highest risks. As a result, this study was done to identify and compare the potential risk associated with various SMR designs. A rigorous risk evaluation can help regulatory agencies set safety criteria for SMRs. This study conducts a comparison of Stage-1 PSA data for the ACP100, NuScale, and BWRX-300 reactors. The research method consisted of gathering and analysing data from available reports and technical documentation for each type of SMR design. Key risk factors and event scenarios are identified and assessed to estimate the potential risks associated with each design. Furthermore, the risk profile of SMRs is compared to typical large-scale nuclear reactors to determine variations in safety. Findings of this study indicates that the ACP100's direct vessel injection line brake contributed 66.3% to the core damage frequency (CDF), the NuScale's loss of the auxiliary power supply system contributed 33.6% to the CDF, and the BWRX-300's boron injection system contributed 95.6% to the CDF. Furthermore, SMR has a lower CDF value per reactor year than typical reactor, which range from 1×10^{-8} to 4.47×10^{-7}, compared to 1×10^{-5} to 1×10^{-4} for conventional reactors. The findings of the study can inform future design choices for SMR.

1. Introduction

A small modular reactor (SMR) is a type of nuclear power facility. According to the International Atomic Energy Agency (IAEA)(1), SMR is an advanced nuclear reactor with a power capacity of up to 300 MW(e) per unit, or one-third of the power capacity of conventional nuclear reactors with capacities of more than 1000 MW(e). Given the benefits offered by SMR, SMR development projects are underway in various nations, including the United States (US), Canada, China, Russia, and the United Kingdom (2).

IOP Conf. Series: Materials Science and Engineering **1326** (2025) 012009 doi:10.1088/1757-899X/1326/1/012009

There are around 80 SMR designs around the world, the majority of which are under development, including four SMRs in Argentina, China, and Russia that are nearing completion (1). The progress levels of these SMRs vary, ranging from early design stages to construction readiness (e.g., ACP100 in China).

The growing interest in SMRs is largely driven by their potential to address climate change and enhance energy security by providing a zero-carbon, reliable energy source. SMRs are uniquely positioned to complement renewable energy technologies and reduce reliance on fossil fuels, aligning with global decarbonization goals (3). The Paris Agreement emphasizes the need for countries to adopt measures that will limit global warming and reduce greenhouse gas emissions. In this framework, SMRs present a unique opportunity to transition from fossil fuel dependency to cleaner energy sources.

The government is considering using nuclear technology as a source of national energy, such as SMR, because it guarantees zero carbon emissions while still meeting the country's energy needs (4). The SMR is expected to address requirements that conventional nuclear reactors cannot fulfil due to its compact size, safe modular construction method, and lower risk of radiation exposure (5).

However, the lack of comprehensive comparative risk assessments for different SMR designs presents a significant challenge for stakeholders. This gap in knowledge hinders informed decision-making by regulators, investors, and the public, as the relative safety and risks of various designs remain unclear. Effective risk assessment methodologies are essential for building public trust, streamlining regulatory processes, and ensuring the safe and successful deployment of SMR technologies (6).

A comprehensive accident risk analysis can assist reveal flaws in the design and operation of SMR, allowing for modifications in strategy and technology to prevent the risk from occurring (7). Understanding and lowering the risk of accidents associated with SMR design is critical for improving safety and obtaining regulatory approval. Effective accident risk management can boost the success of SMR implementation as a dependable and sustainable energy source.

As of the unique technology and system of each SMR design, the sources of accident risk vary. Probability of safety assessment (PSA) Level 1 is a method for assessing the safety and risk of a nuclear power plant that focusses on determining the likelihood of damage to the reactor core caused by external and internal threats. Existing studies primarily focus on the benefits and safety characteristics of a single type of SMR design. The absence of information and data for a full comparative examination of the causes of risk for various SMR designs makes it impossible to determine which design features, systems, and technologies contribute the most to safety while posing the greatest danger.

Addressing this research gap is crucial to enhance the understanding of SMR risks and inform the design of safer and more efficient reactors. By conducting a thorough comparative analysis of accident risks across different SMR designs, stakeholders can better identify and mitigate vulnerabilities, paving the way for widespread adoption of this innovative technology as a cornerstone of global energy solutions.

As a result, this study was done to identify and compare the contributing factors of core accident risk associated with various SMR designs. The safety and efficiency of the SMR design can be improved by using the PSA results to apply safer and more effective nuclear technologies. This research aims to determine the causes for core accident risk in SMR designs, compare those causes to those in conventional reactors, and compare the core damage frequency (CDF) between SMR and conventional reactors.

Therefore, this study selects specific SMR designs for discussion based on several criteria, such as the reliability of the technology for short-term/near-term use, the maturity of the reactor technology design, the capability for electricity generation, and the likelihood of obtaining a reactor construction license. The criteria for the reliability of technology for near-term use focus on reactors that have been developed or are expected to be developed by 2032 and operational by 2035, thereby limiting the selection to light water reactor (LWR)-type due to the availability of their technology. The level of technological maturity is further restricted to SMR models with completed preliminary designs to ensure readiness for near-term development. Additionally, SMRs designed for electricity generation are prioritized over those designed solely for heat generation to meet the growing electricity demand from

IOP Conf. Series: Materials Science and Engineering **1326** (2025) 012009 doi:10.1088/1757-899X/1326/1/012009

consumers. Finally, this study focuses on LWR-SMRs due to their high probability of obtaining construction licenses. This is attributed to their long-standing technological development and the broader and more sustainable knowledge of related legal and safety regulations.

2. Methods

There are 83 SMR models that are either proposed or under development worldwide (8). This study focuses only on some SMR designs due to a selection of considerations, including the technology's reliability for short/medium term use, design maturity, capacity to generate power, and the likelihood of acquiring a construction license. Only three SMR designs have available information on PSA Level 1 for comparing accident risk causes which are ACP100, BWRX-300 and NuScale. Table 1 provides a summary of the key features of the ACP100, NuScale, and BWRX-300 designs. These designs were chosen because they give an extensive foundation for examination. The ACP100's compact design simplifies mechanical systems, but it is primarily reliant on passive cooling, which poses dangers during prolonged loss-of-cooling events. The BWRX-300's streamlined, bigger design minimizes system complexity, but it may struggle to manage heat under extreme temperatures. The modular architecture of NuScale isolates dangers across units, yet the smaller module size necessitates exact integration of passive safety elements. The comparisons of the causes of accident risk, as well as initiating event (IE) leads to core damage frequency (CDF) between three SMR designs, as well as comparison with conventional reactor was made based on the method mentioned earlier in the three main documents (9-11).

Table 1. Summary of the key design features of ACP100, NuScale, and BWRX-300.

Design Features	ACP100	NuScale	BWRX-300
Nuclear Steam Supply System (NSSS)	Integrated design consisting of the reactor core, 16 once-through steam generators (OTSG)	Integrated design consisting of the reactor core, steam generator, and pressurizer within the RPV	Consists of the RPV and fine motion control rod drive (FMCRD)
Fuel Assembly	57	37	240
Reactor Core Height (m)	2.15	2.00	3.80
Fuel Enrichment (%)	1.9 - 4.95	4.95	4.95
Reactivity Control Mechanism	Control rod drive, burnable poison solid rods, and boron	Control rod drive, boron	FMCRD and burnable poison solid rods
Reactor Cooling System (RCS)	Forced circulation	Natural circulation	Natural circulation
Steam Generator Location	16 OTSGs attached to the RPV	Annular space between hot leg riser and RPV	No steam generator needed
Pressurizer	Located outside the reactor vessel	Pressurizer within RCS pressure control	No pressurizer
Main Pump	Pumps attached to the RPV	No pumps required	No pumps required
References	(9)	(10)	(11)

3. Results and discussion

3.1. Identification of core accident risk causes for SMR design based on PSA level 1.

Based on Table 2, the PSA Level 1 data reveal the CDF for ACP100, NuScale, and BWRX-300 are respectively 4.47×10^{-7} (9), 1×10^{-8} (10) and 1.23×10^{-7} (11) per reactor year, respectively. Direct injection vessel damage (DVI) accounts for 66.34% of ACP100's CDF risk, followed by moderate loss of coolant (LOCA) (9.37%), pressure vessel rupture (7.04%), and steam tube damage (6.25%) (9).

Furthermore, the loss of NuScale's support system had a significant impact on CDF of up to 33.6%, followed by loss of external power, reactor general shutdown, RCS failure causing LOCA in containment, LOCA injection line failure, accidental opening of emergency core cooling system valves, and DC power loss, all of which contributed 27.1%, 19.9%, 9.2%, 4.3%, 4.2%, and 1.8%, respectively (10).

Transients involving the evaporative isolation system (ICS) contribute the most to the CDF for BWRX-300, accounting for up to 95.31%, followed by transients involving the boron injection system (BIS) (0.31%), LOCA (0.14%), and vessel rupture (0.08%) (11). Table 2 shows the main risk factor (CDF) and components or systems that have a substantial impact on the core accident risk for each design.

Table 2. Risk factors for each SMR design.

SMR design	Core damage frequency, CDF (per reactor year)	Initiating event (IE)	Contribution of IE to CDF (%)	References
ACP100	4.47×10^{-7}	DVI line break	66.34	(9)
		Middle LOCA	9.37	
		Rupture of pressure vessel	7.04	
		Steam tube break	6.25	
NuScale	1×10^{-8}	Loss of support system	33.60	(10)
		Loss of offsite power	27.1	
		General reactor trip	19.9	
		Reactor coolant system LOCA inside containment	9.2	
		Chemical and volume control system LOCA injection line inside containment	4.3	
		Spurious opening of an emergency core cooling system valve	4.2	
		Loss of DC power	1.8	
BWRX-300	1.23×10^{-7}	Transient involving ICS	95.31	(11)
		Transient involving BIS	0.31	
		LOCA	0.14	
		Vessel rupture	0.08	

According to the findings of Deng et al. 2020 (9), DVI line damage is the most significant initiating event (IE) in ACP100, accounting for 66.3% of the overall CDF. This demonstrates how crucial the DVI line is to the reactor's safety and risk profile. The DVI channel is a critical component that injects emergency coolant directly into the reactor core during an accident like LOCA. This direct coolant injection is critical for sustaining core cooling and avoiding excessive heat that could damage the reactor core (12). The DVI line breakdown was the most severe and difficult LOCA accident for the ACP100 because of its design, which merges crucial components in a single reactor vessel.

The loss of the support system (LOSS) is the largest contributor to NuScale's CDF, which is up to 33.6%. According to the IAEA (2020) (13), the support system is a critical supplemental system for nuclear reactors that provides needs or services such as electrical power, cooling, and equipment to enable safe operation. The failure of this system can cause a series of events that endanger the reactor's safety while increasing the risk of damage to the reactor core. For example, the loss of AC power to critical systems can cause a series of failures that impair the reactor's ability to cool and maintain heat during operation (10).

Next, IE such as transients induced by isolation condenser system component failures, account for 95.31% of CDF BWRX-300 contributions (11). The ICS is a passive safety device that automatically removes decay heat from the reactor core via natural circulation in the heat exchanger (14). The BWRX-300 relies on passive safety mechanisms like ICS to regulate decay heat and keep the core cool in the case of a crash. The failure of all ICS units that occur simultaneously leads the reactor to lose the principal mechanism for heat removal while also increasing the possibility of damage to the reactor core.

3.2. A comparative analysis of risk causes and CDF for SMR and conventional reactors.

The following section compares the risk profile of SMR (ACP100, NuScale, BWRX-300) with conventional reactors. Table 3 compares the risk profile of SMR and traditional reactors, with SMR having a lower CDF value than conventional reactors. NuScale's safety and passive system design result in the lowest CDF of 1×10^{-8} per reactor year (10). Transient is the most prominent IE group in terms of contribution to CDF. This is evident from the fact that the bulk of IEs that contribute to CDF include support system failures, ICS transients, and feedwater losses. The second contributor is from the category of LOCA events, namely DVI line damage. The contribution to CDF of 95.62% (11) for the BWRX-300 demonstrates that the ICS system requires the most attention because practically the whole core risk is associated with its failure. Furthermore, DVI line damage as a significant factor emphasizes the significance of preserving the reactor's emergency cooling system. NuScale's risk profile is more balanced, with LOSS occurrences accounting for one-third of the overall CDF, demonstrating that, while LOSS is a large contributor, the reactor's design and safety systems contain extra safety measures to assist mitigate the impact of risk. In comparison, the conventional reactor shows the least contribution of IE to CDF (12%) (15) due to loss of feedwater compared to SMRs (Figure 1).

Table 3. Comparison of the risk profile of SMR with conventional reactors.

Parameter	ACP100	NuScale	BWRX-300	Conventional reactor
CDF (per reactor year)	4.47×10^{-7}	1×10^{-8}	1.23×10^{-7}	1×10^{-5} to 1×10^{-4}
Significant IE	DVI line break	LOSS	Transient involving ICS	Loss of feedwater
Contribution to CDF (%)	66.34	33.60	95.31	12.00
References	(9)	(10)	(11)	(15)

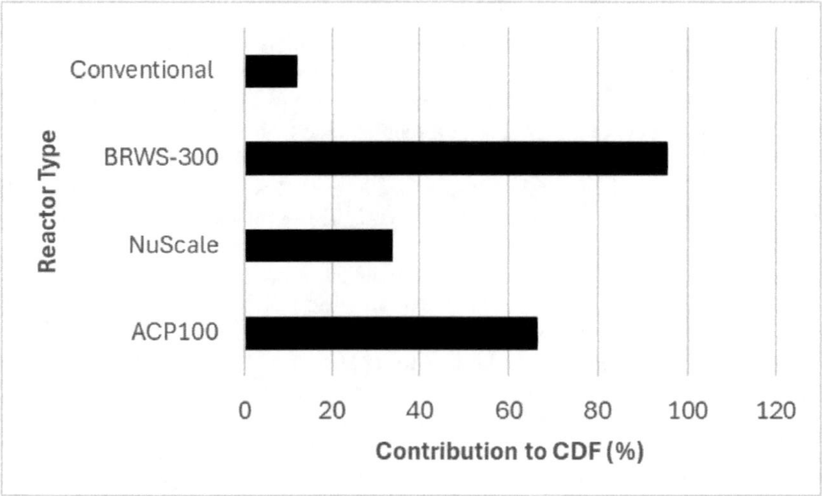

Figure 1. Comparison contribution of IE to CDF between APC100 (9), NuScale (10), BRWS-300 (11) and conventional reactor (15).

3.3. Evaluating risk factors and mitigation strategies in SMR designs.

The analysis of CDF and the contribution of various IE across three small modular reactor (SMR) designs—ACP100, NuScale, and BWRX-300—provides important insights into their safety profiles. These findings illustrate the specific problems and risks associated with each reactor type, as well as the efficacy of their safety systems.

The DVI line damage in ACP100 accounts for 66.34% (9) of the total CDF, highlighting its vital importance in reactor safety. The DVI line failure has a substantial impact on the emergency cooling function, which is critical in case of a LOCA. Enhancements to the DVI system's structural integrity and redundancy could decrease this danger, assuring sustained core cooling in an emergency. The potential implications of a DVI line failure, including delayed coolant injection and increased thermal load on the reactor core, highlight the importance of rigorous testing and maintenance procedures (12).

The NuScale reactors require support systems is demonstrated by LOSS providing 33.6% of its CDF (10). Failure in these systems can lead to serious safety issues, as seen by the loss of AC power, which affects heat removal and reactor stability (16). To address this, enhancements passive safety measures and improved backup systems should be considered (17). Furthermore, NuScale's balanced risk profile when compared to other SMRs demonstrates the efficacy of its inherent safety design. Nonetheless, possible gaps in backup power sources, as well as the risk amplification that comes with them, require ongoing monitoring.

The BWRX-300 is highly dependent on the passive functionality of the ICS, with transients in this system contributing a striking 95.31% to its CDF (11). This dependency raises concerns about the system's reliability during simultaneous failures, where decay heat removal becomes critically compromised (18). Regular testing of the ICS's passive safety mechanisms and the integration of supplementary heat removal methods could address these vulnerabilities (19). The impact of ICS transients on reactor safety underscores the importance of designing multi-layered defences.

4. Conclusion

To conclude, the significant cause of accident risk for ACP100, NuScale, and BWRX-300 is DVI line damage, LOSS, and transients related to ICS, which respectively contribute as much as 66.34%, 33.6%, and 95.31% to the total amount of CDF. Comparisons with conventional reactors reveal that transient events are the leading contributors to accident risks across all reactor designs, followed by LOCA events. Transients significantly impact the CDF as they demand immediate and complex safety system responses to prevent core damage. Importantly, the lower CDF range of SMRs which are in the range of 1×10^{-8} to 4.47×10^{-7} per reactor year compared to conventional reactors 1×10^{-5} to 1×10^{-4} per reactor year demonstrates their enhanced safety and advanced design features. Furthermore, the major impact of transient events on CDF emphasises the importance of having strong safety systems and timely mitigation techniques in reactor operations. While this study gives useful insights into accident risks and safety features, it only examined three SMR designs and performed PSA Level 1 analysis. To improve on these findings, future research should incorporate more SMR designs, perform PSA Level 2 and Level 3 evaluations, and investigate a broader range of disaster scenarios. These additional initiatives would not only address new issues in transitory management and LOCA event reduction but would also provide a more thorough understanding of accident risks, safety systems, and mitigation solutions. Such developments are necessary to assure the safe and reliable operation of future SMRs technologies.

5. References

1. IAEA. Small modular reactors: a new nuclear energy paradigm. 2024.
2. Murakami T, Anbumozhi V. Global situation of small modular reactor development and deployment. Senayan; 2021.
3. U.S. Department of State. Response to climate change and energy security: opportunities and challenges for SMRs. 2023.
4. Bernama. Malaysia keeps "no nuclear policy" stance on power generation. 2023.
5. WNA. Small nuclear power reactor 2024 [Available from: https://world-nuclear.org/information-library/nuclear-fuel-cycle/nuclear-power-reactors/small-nuclear-power-reactors.
6. NEA. Small modular reactors: challenges and opportunities. 2021.
7. Smith C, Prescott S, Koonce T. Advanced small modular reactor (SMR) probabilistic risk assessment (PRA) demonstration Idaho National Loboratory; 2014.
8. IAEA. Advances in small modular reactor technology developments. Austria; 2022.
9. Deng J, Xu Y, Qiu Z, Zhang H, Wu L, Wang X, et al. Research on level 1 PSA and risk-informed design for ACP100. Ann Nucl Energy. 2020;144(3):107542.
10. NuScale Power LLC. Probabilistic risk assessment and severe accident evaluation. 2023.
11. Trundle G. Reliability assessment of passive ICS in an SMR as part of the PSA analysis. Stockholm: KTH Royal Institute of Technology; 2023.
12. Ekariansyah AS, Widodo S. Performance analysis of AP1000 passive systems during direct vessel injection (DVI) line break. Atom Indonesia. 2016;42(2):79-88.
13. IAEA. Advances in small modular reactor technology developments. Austria; 2020.
14. IAEA. Passive safety systems and natural circulation in water cooled nuclear power plants. Vienna; 2009.
15. Apostolakis GE. Reactor safety: the emergence of probabilistic risk assessment. Elements of reactor design, operations, and safety. Department of Nuclear Science and Engineering2006.
16. Welter K, Reyes Jr JN, Brigantic A. Unique safety features and licensing requirements of the NuScale small modular reactor. Front Energy Res. 2023;11.
17. Hosseini SA, Akbari R, Shirani AS, Auria FD. Investigation of the natural circulation flowmap in NuScale small modular reactor. International Journal of Reliability, Risk and Safety: Theory and Application. 2021;4(1):95-100.

IOP Conf. Series: Materials Science and Engineering **1326** (2025) 012009 doi:10.1088/1757-899X/1326/1/012009

18. Kitou K, Ishida N, Tamura A, Ishibashi R, Kanada M, Kamoshida M. Plant system and required water pool capacity for large scale BWRs with inherently safe technologies. Mechanical Engineering Journal. 2015;2(5):1-8.
19. Papini D, Cammi A. Modelling of heat transfer phenomena for vertical and horizontal configurations of in-pool condensers and comparison with experimental findings. Science and Technology of Nuclear Installations. 2010.

Acknowledgements

Acknowledgment is given to the Nuclear Technology Research Centre, UKM for the support of this study.

IOP Conf. Series: Materials Science and Engineering **1326** (2025) 012010 doi:10.1088/1757-899X/1326/1/012010

Characterizing radioactive source casings for nuclear forensics using a multi-technique laboratory approach

Zalina Laili[1, a], Suzilawati Muhd Sarowi[1], Nazran Harun[1], Nadira Kamarudin[1], Susan Maria Sipaun[1], Hishamudin Husin[1], Muhamad Azfar Azman[1] and Muhamad Rawi Zain[1]

[1]Malaysian Nuclear Agency (Nuclear Malaysia),
Bangi, 43000 Kajang, Selangor Darul Ehsan
MALAYSIA

[a]liena@nm.gov.my

Abstract. This investigation characterised various types of radioactive source casings for nuclear forensic applications using a multi-technique laboratory approach. Quantitative and qualitative assessments were conducted, including visual inspection and physical characterisation, with advanced tools such as Scanning Electron Microscopy (SEM), and X-ray imaging. The results demonstrated that these methods provided valuable information on the materials, corrosion, wear, and physical damage of radioactive source casings. This multidisciplinary approach enhanced the effectiveness and quality of forensic investigations, supporting the identification and prevention of illicit trafficking of radioactive materials.

1. Introduction

In Malaysia, radioactive sources are extensively used in various industrial applications, including measuring material properties such as thickness and density, conducting non-destructive tests in construction, and controlling manufacturing processes. A common industrial application of these sources is the nuclear gauge, a type of industrial radiography. However, on August 10, 2018, an incident involving a missing radioactive source commonly used in industrial radiography alarmed the public [1]. The device, containing the radioactive isotope Iridium-192 (^{192}Ir), is utilized for testing, calibration, and inspection across industries such as oil and gas, as well as other heavy sectors. Following the incident, concerns have grown among Malaysian authorities about the potential use of this material in a terrorist attack. When radioactive sources are lost or stolen, they pose a serious danger to both human safety and the environment if they fall into unauthorized hands. This material could be used in the construction of a dirty bomb or radiological dispersal device (RDD) when combined with conventional explosives like dynamite.

According to records from the Department of Atomic Energy Malaysia (Atom Malaysia), at least 18 cases of lost radioactive sources were reported between 2005 and 2019 [2]. Additionally, some radioactive sources lack adequate control, accountability, and disposal protocols. Orphaned radioactive sources or other radioactive materials are sometimes inadvertently collected as scrap metal meant for recycling. This has heightened the Malaysian government's concern regarding these orphaned sources [2]. When nuclear or radioactive material is discovered outside regulatory control, nuclear forensic investigation techniques can be employed. Nuclear forensics is a scientific discipline focused on investigating and obtaining information about the history and origin of nuclear and radioactive materials. It plays a crucial role in national and international response plans for incidents involving such materials that are outside regulatory control. These investigations are instrumental in preventing nuclear terrorism, ensuring the security of nuclear facilities, and upholding international nuclear non-proliferation agreements.

In nuclear forensics, the concept of signatures refers to material features or characteristics such as isotopic ratios, elemental compositions, physical attributes, and chemical properties. These characteristics enable the material to be associated with specific individuals, locations, and processes [3]. The physical features of radioactive sources provide valuable information for distinguishing between different sources. These characteristics include aspects such as size, shape, number of capsules, casing types, manufacturer details, dimensions, welding specifics, placement of spacers, engravings, and materials used in construction. Therefore, the materials used for radioactive source casings must be stable and inert to prevent structural degradation and reactions with aggressive substances in a radiation field. While stainless steel is the most common material used, alternatives such as aluminium, titanium, tantalum, brass, and zircaloy are also utilized. The specific alloy or type of metal is important for forensic purposes as it can be a crucial part of a set of characteristics that constitute a signature [4]. Another obvious physical signature is the label on the radioactive source casing. The label may contain information such as the logo, manufacturer's abbreviation, isotope name, activity, source serial number, model number, date of manufacture, and the word "radioactive." The amount of information on the radioactive source casing depends on the physical size of the source itself.

The comprehensive characterisation of radioactive source casings is a critical aspect of nuclear forensics, as it enables the identification and understanding of these materials' properties, origins, and potential applications. Identifying the signatures of radioactive source casings requires the use of diverse techniques. The selection of a method to characterize a radioactive source casing depends on the specific parameters to be evaluated, whether they relate to physical, chemical, or radiological aspects. However, despite its significance, challenges remain in achieving thorough and accurate characterisation due to the complex nature of radioactive sources and the diversity of materials used in their casings. This study addresses the pressing need for a robust and systematic approach to characterising radioactive source casings, aiming to overcome existing limitations and enhance the capabilities of nuclear forensics in Malaysia.

By employing a multi-technique laboratory approach, this study aims to uncover the physical and chemical properties of radioactive source casings, facilitating their identification, differentiation, and attribution. The primary objective is to develop a comprehensive methodology for characterising these casings, utilising a combination of advanced analytical techniques and methodologies. Through systematic experimentation and analysis, the study seeks to establish a reliable framework for assessing the material composition, structural integrity, and potential forensic signatures of these casings.

2.Method and materials

2.1. Sample Selection Criteria, Sampling and Sample Collection

The radioactive source casings were collected from the Waste Technology Development Centre (WasTeC) of the Malaysian Nuclear Agency (Fig. 1). For the selection of radioactive source casings, a diverse sample was chosen from different sources, including various isotopes such as Cesium-137(^{137}Cs), Cobolt-60 (^{60}Co), and Americium-241 (^{241}Am), and casings made by different manufacturers. This selection strategy ensures coverage of a wide range of casings in use, considering differences in design, materials, and manufacturing processes. Priority is given to samples that align with the investigation goals, considering factors such as accessibility and practicality. Additionally, methods such as random or systematic sampling were employed to accurately represent the available range of casings.

(a) Sample 1 : Nuclear Density Gauge Casing. Designed to Store: ^{137}Cs. Origin: Commonly used in construction and civil engineering

(b) Sample 2: Nuclear Density Gauge Casing. Designed to Store :^{137}Cs. Origin: Commonly used in construction and geotechnical engineering.

(c) Sample 3: Radiometric Density Source Casing. Designed to Store: ^{60}Co. Origin: Commonly used in industries.

(d) Sample 4: Lead Pot Container. Designed to Store: ^{60}Co. Origin: Commonly used in portable gamma-ray devices.

(e) Sample 5: Lead Pot Container. Designed to Store :^{137}Cs. Origin: Commonly used in radiometric applications and industrial gauging.

(f) Sample 6: Nuclear Moisture Density Gauge Casing. Designed to Store: ^{241}Am-241/Be. Origin: Commonly used in construction and geotechnical engineering.

Figure 1. Types of Radioactive Source Casings for Characterisation Study: Sample 1 (a), Sample 2 (b) – Nuclear Density Gauge Casings, Sample 3 (c) – Radiometric Density Source Casing, Samples 4 (d) and 5 (e) – Lead Pot Containers, and Sample 6 (f) – Nuclear Moisture Density Gauge Casing.

2.2. Material Characterisation Strategies and Techniques

Several factors are considered when selecting samples of radioactive source casings for material characterisation. The choice of casings depends on the selected analysis techniques and the overall purpose of the study.

2.2.1Physical Characterisation

Physical characterisation methods for radioactive source casings include visual inspection, size, shape, surface texture, photography, weight determination, and dimensioning [5]. Standard laboratory tools such as a caliper (Mitutoyo model CD-15APX), weighing bench scale (Excell model FTHW3-150), and a digital camera (Nikon D600) were used for detailed visual inspection. External surfaces were carefully examined under appropriate lighting and magnification to detect signs of damage, markings, or unusual features indicating unauthorized access. Surface contamination was measured using a survey meter (Red Eye B20/B20-ER), which provided quantitative readings of radioactive contamination levels on the casing surfaces

Material composition and construction details were analyzed to document any unique modifications. Photography was crucial for documenting both external features and, when feasible, internal aspects of the casings, capturing close-up shots and using scale references for precise measurements. Physical dimensions and weights were measured and recorded with calibrated tools, and all findings were carefully documented. This thorough physical characterisation is essential for nuclear forensic investigations.

2.2.2 Material Composition Analysis

Surface samples were collected from radioactive source casings to analyse potential surface treatments, coatings, or contamination. A clean, sterile scraping tool was used to gently scrape the casing surfaces, focusing on areas where coatings or treatments were visible. The collected material was placed in sample containers, carefully labelled, and securely sealed to preserve integrity.

The samples were analysed using a scanning electron microscope (SEM-EDX) (model FEI 400) to examine their microstructure and elemental composition, providing a detailed understanding of the material's characteristics. Scanning Electron Microscopy (SEM) with Energy Dispersive X-ray (EDX) spectroscopy is ideal for examining the surface morphology and composition at high magnifications. SEM provides detailed images of the sample's surface topography, which is essential for observing fine structural features such as the flake-like structures, wear patterns, or corrosion products present.

2.2.3 X-Ray Imaging

This imaging setup utilised a portable x-ray system XRS4, with adjustable pulse settings and a high-resolution 10" x 12" digital x-ray detector SENTINEL FXG-DR1012. The pulsed x-ray generator produces a maximum of 370 kV x-rays, capable of penetrating dense materials up to 3.81 cm (1.5 inch) of steel [6]. The pulse count was set between 30 and 40 pulses and time delay was set 10 seconds. The x-ray-to-detector distance is 1.0 meter. The detector detects x-ray radiation within the energy range of 40 to 330 kVp [7]. Imaging parameters were adjusted according to the material composition and thickness of the casings to achieve high-resolution images. The casings were carefully aligned with the x-ray source and detector, ensuring the x-ray beam passed directly through the sample for optimal detection. Image acquisition dan analysis was performed using MultiVision HD software. Two samples of returned radioactive containers and equipment, an empty lead pot container (sample 4) and nuclear moisture density gauge (sample 6), were investigated. Both the lead pot and the gauge are clear of any radioactive material residue at the time of imaging. In typical use, the gauge employs a small ^{241}Am/Be neutron source (1.85 GBq, 50 mCi) and a ^{137}Cs gamma source (0.37 GBq, 10 mCi) [8]. The neutron source is located in the centre of a lead cup, which is screwed into the shielded area at the centre of the gauge base. The ^{137}Cs source is located within the end of a source rod, which is housed within the guide tube. The source is securely fixed to the tip of the source rod and is surrounded by lead and tungsten shielding. The assembly of this gauge includes two units of Geiger-Muller detectors and a single Helium-3 (^{3}He) tube which are used to detect gamma ray and neutron respectively. In this work, both sources have been removed prior to imaging. X-ray imaging was chosen due to its non-destructive nature and ability to provide detailed insights into the internal structure of materials. This technique allows for the penetration of the sample, revealing internal features such as cracks, phase changes, and structural integrity. X-ray imaging is especially useful for analysing larger or more complex samples, as it provides a comprehensive evaluation of the material's bulk properties without the need for sample alteration.

3. Result and discussions

3.1. Physical characteristic of radioactive source casings

The physical characterisation results are presented in Table 1. The physical characterisation of the containers reveals a variety of shapes, surface textures, and conditions, all of which point to significant wear and degradation. The cylindrical containers, ranging in size from smaller units of around 8.3 cm in width and 16.5 cm in height to larger ones reaching 42 cm in height, display rough and worn surfaces. Many have peeling paint, corrosion, and, in some cases, exposed metal due to flaking. Key identifiers such as radioactive material labels and metal-engraved details, which include information on isotope type, activity, and measurement dates, are visible but show signs of wear. Corrosion is present on multiple surfaces, particularly around the base and fastenings, which could pose a risk to structural integrity and safety. The cylindrical shape is not only practical for sealing the source with a seamless weld, but it also allows radioactive material, often in the form of pellets or powder, to be evenly distributed within the capsule through shaking and compression [9].

In the context of nuclear forensic analysis, if such cylindrical containers were found out of regulatory control, this design feature could provide valuable clues. The even distribution of material inside the casing could help experts determine whether the contents are intact or have been tampered with. Additionally, damage to the casing such as signs of excessive shaking or wear could indicate improper handling or attempts to alter the material. Spherical containers, known for their compact and robust designs, are also heavily affected by corrosion, rust, and dirt accumulation. Their metallic surfaces are uneven, and scratches and tool marks suggest frequent handling and maintenance. Many of these containers have worn or missing labels, raising concerns about the clarity and longevity of vital safety information, particularly radioactive material markings.

The rectangular containers, which are designed with attached cylindrical rods, present smoother surfaces than the spherical and cylindrical units, though they too show signs of scratches, faded paint, and wear around mounting points. Despite the less severe surface wear, these containers still exhibit structural wear at points where components like the rods are attached. Overall, corrosion and rust are prevalent across all types of containers, indicating prolonged exposure to environmental conditions and potential lapses in regular maintenance.

The extensive surface wear, fading labels, and tool marks suggest that these containers are frequently handled, but likely without adequate care or upkeep. This level of degradation, particularly around seals and fastening points, could compromise the safety and integrity of these containers, which is especially critical for units storing or transporting radioactive materials. The worn labels and degraded identifiers also present a significant issue for maintaining proper tracking and safety protocols, underscoring the need for improved labelling materials and maintenance routines. This extensive degradation, including worn labels and degraded identifiers, challenges the integrity and safety of the containers and poses difficulties for forensic investigations. Addressing these issues highlights the need for better labelling materials, maintenance routines, and complementary forensic techniques to overcome the limitations imposed by the containers degraded physical state.

IOP Conf. Series: Materials Science and Engineering **1326** (2025) 012010 doi:10.1088/1757-899X/1326/1/012010

Table 1. Physical characteristics of radioactive source casings

Sample	Dimensions (cm)	Weight (kg)	Surface Contamination Reading (cps)	Shape Description	Surface Texture	Unique Identifiers	Signs of Damage/Markings	Photographic Details
1	• Width: 8.3 cm • Height: 16.5 cm • Bottom Diameter: 12.46 cm	10.42	1.30	• Cylindrical body with a circular base. • Flat, round lid on top. • Base has four mounting holes, suggesting it can be fixed to a surface.	• Rough, chipped paint on the surface, showing signs of wear. • Green colour, though some areas are flaking or corroded, exposing metal underneath.	• Marking Number (written in blue marker pen). • Radioactive Material Label • Metal-Engraved Details	• Corrosion: Observed on multiple casing surfaces (right, left, rear, bottom). • Surface Condition: Uneven with roughness; green coating peeling due to rust. • Debris inside/around the container. • Tool marks, and scratches on seals/fastenings. • Labelling: Visible radioactive symbol sticker; • Metal-engraved details (activity, isotope type, measurement date).	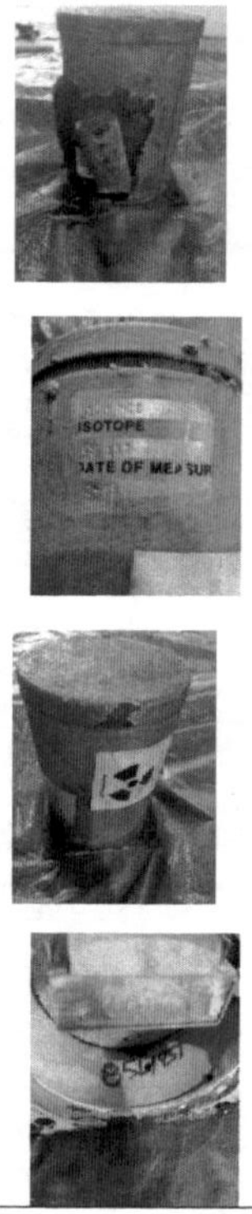

IOP Conf. Series: Materials Science and Engineering **1326** (2025) 012010 doi:10.1088/1757-899X/1326/1/012010

Sample	Dimensions (cm)	Weight (kg)	Surface Contamination Reading (cps)	Shape Description	Surface Texture	Unique Identifiers	Signs of Damage/Markings	Photographic Details
2	• Height: 36 cm • Bottom Diameter: 22 cm	40.07	1.31	• Spherical body • Flat circular base with bolt holes • Cylindrical top for source attachment • Side handle • Compact and robust design	• Rough and corroded • Uneven patches • Discoloration and rust • Scratches visible • Worn labels/symbols • Metallic, weathered surface	• Engraved metal plate with specific details such as activity, isotope type, and measurement date. • Radioactive Material Label attached to the surface.	• Corrosion on multiple surfaces (right, left, rear, bottom). • Scratches on the body surface. • Debris inside/around the container. • Tool marks and scratches on seals/fastenings. • Worn labels and radioactive symbol. • Missing radioactive label.	 
3	• Height: 18 cm • Bottom Diameter: 7.7 cm	21.04	1.3	• Spherical body with flat base. • Metal latch on top	• Dull, metallic, worn with dirt. • Some corrosion.	• Metal latch.	• Scuffs, scratches, and rust on latch.	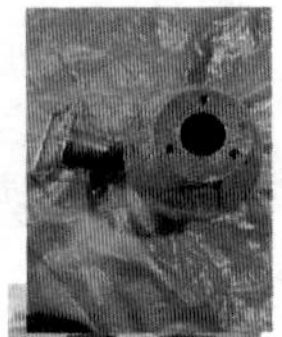 

IOP Conf. Series: Materials Science and Engineering **1326** (2025) 012010 doi:10.1088/1757-899X/1326/1/012010

Sample	Dimensions (cm)	Weight (kg)	Surface Contamination Reading (cps)	Shape Description	Surface Texture	Unique Identifiers	Signs of Damage/Markings	Photographic Details
4	• Width: 18 cm • Height: 15 cm • Length: 29 cm	14.5	1.30	• Rectangular body with a vertical cylindrical rod.	• Smooth • Paint colours: Orange (body), Black (rod) • Scratches on multiple surfaces	• Markings on the control panel.	• Scratches, worn paint, possible signs of tampering near the rod top.	  
5	• Height: 8.4 cm (with lid) 5.3 cm (without lid) • Bottom Diameter: 7.8 cm	10.43	1.30	• Cylindrical with a flat top and bottom.	• Rough, worn, with dirt and signs of oxidation.	• Yellow label partially intact, faint markings.	• Significant wear, dirt, rust spots, and scratches visible on the surface.	

IOP Conf. Series: Materials Science and Engineering **1326** (2025) 012010 doi:10.1088/1757-899X/1326/1/012010

Sample	Dimensions (cm)	Weight (kg)	Surface Contamination Reading (cps)	Shape Description	Surface Texture	Unique Identifiers	Signs of Damage/Markings	Photographic Details
6	• Height: 42 cm (with lid) 37 cm (without lid) • Bottom Diameter: 9.69 cm	33.04	1.31	• Cylindrical container with a flat top	• Smooth, white painted surface with some scratches and markings.	• Red label for radioactive material ("LOADED"). • Radiation warning symbol. • Isotope and activity level listed. • Surface radiation dose information. • Handling instructions and sign-off by responsible personnel. • Origin and destination fields for tracking.	• Surface wear, discoloration, and visible dirt marks.	 

3.2. Morphology Characteristics

SEM plays a crucial role in the forensic analysis of radioactive source casings by providing high-resolution imaging and detailed surface characterisation. Figure 2 presents SEM images of Sample 1: (a) shows an SEM image of metal powder obtained from scraping the sample at 2000X magnification, and (b) displays the corresponding EDX results. The SEM image reveals a petal-shaped hierarchical pattern that varies in size and scale, potentially indicating specific microstructural or surface features. Agglomerates of metal particles are also visible in the SEM image. These agglomerates could consist of metal particles from the outer surface, along with possible contaminants or oxidation products. The EDX results revealed a high oxygen content, suggesting that the sample may have undergone oxidation, indicative of corrosion. Corrosion occurs when metals react with oxygen in the environment, forming oxides or other corrosion products on the surface [10]. Elevated oxygen levels strongly suggest that the material was exposed to the environment, leading to oxide formation consistent with corrosion. Additionally, the EDX showed iron (Fe) as the dominant element, indicating a steel or iron-based alloy. The presence of titanium and aluminium could be from coatings or surface treatments, while platinum peaks may indicate platinum-based coatings used during analysis.

Figure 2 (c) is SEM image of metal powder obtained from scraping the Sample 2 at 2000X magnification, and (d) displays the corresponding EDX results. The SEM image of Sample 2 revealed an irregular, rough surface with clearly visible flake-like structures and particles distributed across the sample. The surface appears to show signs of material fragmentation and flaking, which could be the result of corrosion or wear. This type of deterioration is common in materials exposed to harsh environmental conditions, such as the presence of radiation, moisture, or other degrading factors. For instance, a study on metallic powders handling and storage conditions found that repeated environmental exposure can lead to inconsistent flowability and moisture content changes, which may affect powder properties and behavior [11]. Oxidation layers could also form on the surface, breaking off into small particles as the material degrades. The SEM image captures these layers, suggesting that the metal has been exposed to oxidative environmental conditions.

Figure 2 (e) is SEM image of metal powder obtained from scraping the Sample 3 at 2000X magnification, and (f) displays the corresponding EDX results. The particles appear to have a variety of shapes, with some showing more angular, sharp edges, while others are more rounded. This diversity in shape suggests different mechanisms of particle formation, possibly including mechanical scraping and corrosion. This observation is supported by the EDX analysis, which identifies high concentrations of carbon and oxygen, indicating the presence of organic contamination or oxidation, potentially in the form of lead oxides (PbO). Additionally, trace elements such as sodium, silicon, sulphur, zirconium, and titanium were detected, which could signify contamination or minor alloy components. The presence of platinum is likely due to sample coating for SEM imaging. The combination of the SEM and EDX results highlights surface degradation and contamination, which may affect the integrity and long-term performance of the lead pot in nuclear applications. For example, a study on the corrosion-erosion behavior of steel exposed to flowing Pb-17Li (a lead-lithium eutectic) demonstrated that small-scale corrosion-erosion was caused by the removal of small protrusions from the steel surface. This degradation was observed through SEM imaging, highlighting the importance of surface integrity in such applications [12].

Figure 2 (g) is an SEM image of metal powder obtained from scraping Sample 4 at 2000X magnification, and (h) displays the corresponding EDX results. The SEM image reveals an irregular, rough surface with distinct flake-like structures and dispersed particles, indicative of material degradation processes such as fragmentation and flaking. These surface features likely result from wear, corrosion, or environmental exposure, including radiation, moisture, or oxidative conditions [13]. The corresponding EDX spectrum confirms the elemental composition of the sample, showing significant peaks for elements such as carbon (C), oxygen (O), sulfur (S), sodium (Na), aluminum (Al), titanium (Ti), zirconium (Zr), and lead (Pb). The presence of oxygen and sulfur suggests oxidative and sulfide layer formation on the metal's surface, likely contributing to the observed material flaking.

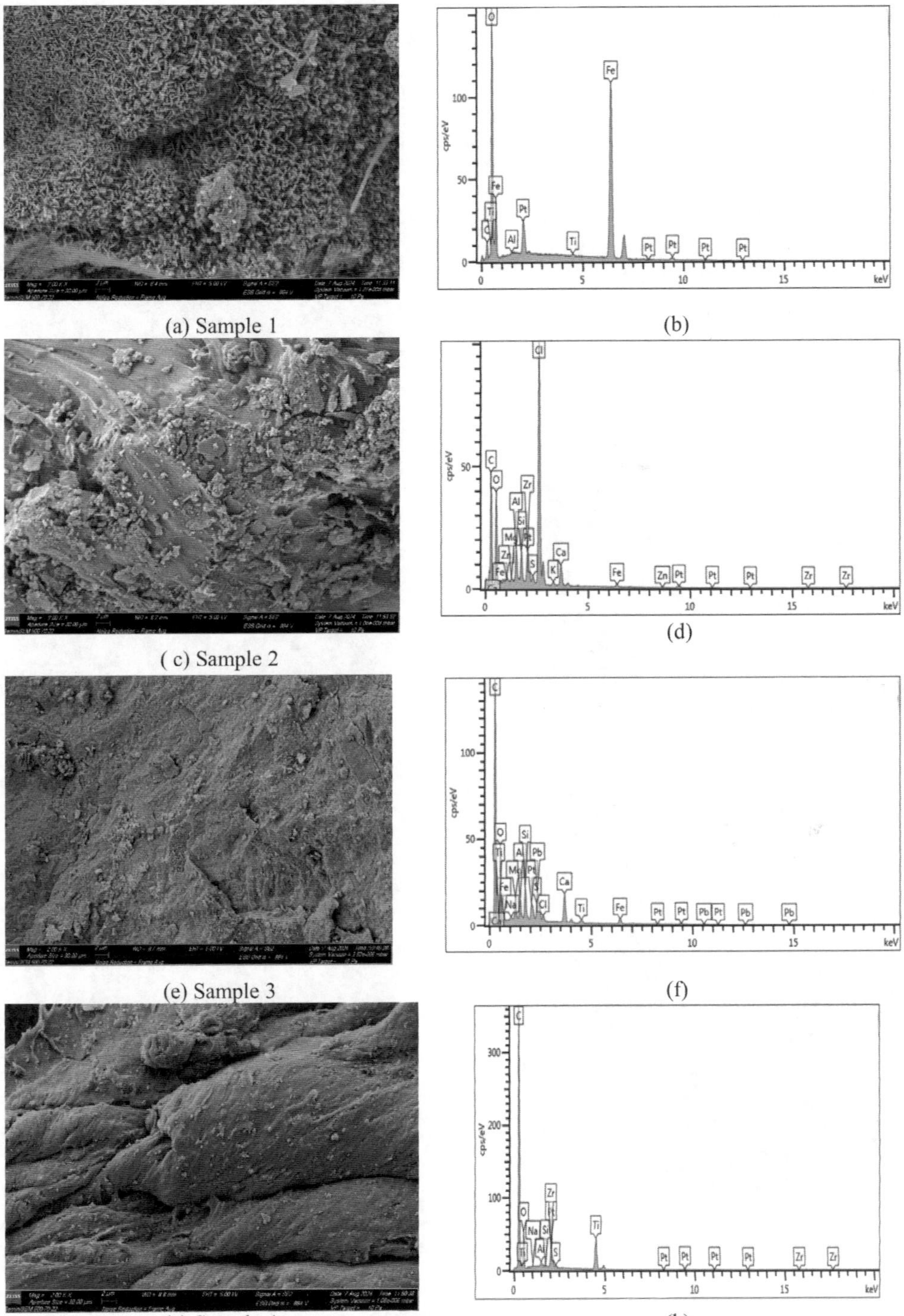

(a) Sample 1 (b)
(c) Sample 2 (d)
(e) Sample 3 (f)
(g) Sample 4 (h)

Figure 2. SEM Images and EDX Results of Metal Powder Obtained from Scraping Samples 1- 4.

IOP Conf. Series: Materials Science and Engineering **1326** (2025) 012010 doi:10.1088/1757-899X/1326/1/012010

3.3. X-Ray Imaging

X-ray radiography is a non-invasive evaluation technique that offers an efficient alternative to destructive methods that involves cutting and drilling. It is widely used to detect defects, cracks, and corrosion in industrial components, however, it is also useful in material testing and inspection. By employing x-ray imaging, researchers can obtain detailed information into the internal structure and composition of objects. X-ray images reveal variations in material density or thickness, allowing for an assessment of an object's structural integrity. This non-destructive method eliminates the need for disassembly, preserving the object's original state, allowing forensic investigation. Figure 3 show the x-ray images for the lead pot container. The lead shielding has a maximum thickness of 4 centimeters, thus the x-ray is mostly attenuated. Internal defects with the lead structure such as cracks or deformation is not easily distinguishable. However, the overall shape is identifiable in the images and is consistent with the sample features. The empty slot (shown in Fig. 3(c)) which is for the placement of a source pen within the pot container is not clearly distinguished but rather shows up as the bright area in images (Figs. 3(a) & (b)). The images suggest that the lead material is intact and the container may be re-used or repurposed.

(a) Top-cap

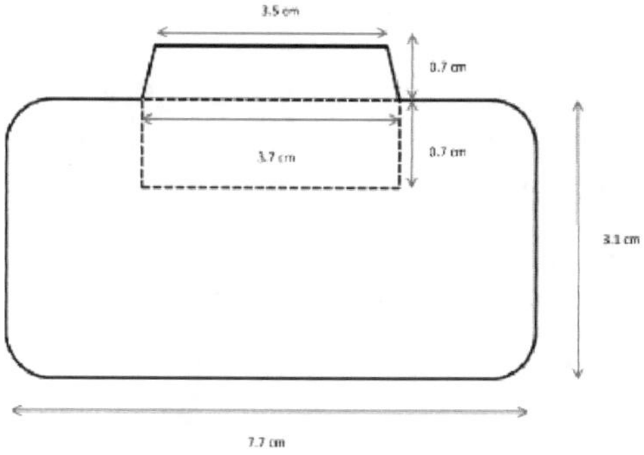

(b) Lead pot (cap and body)

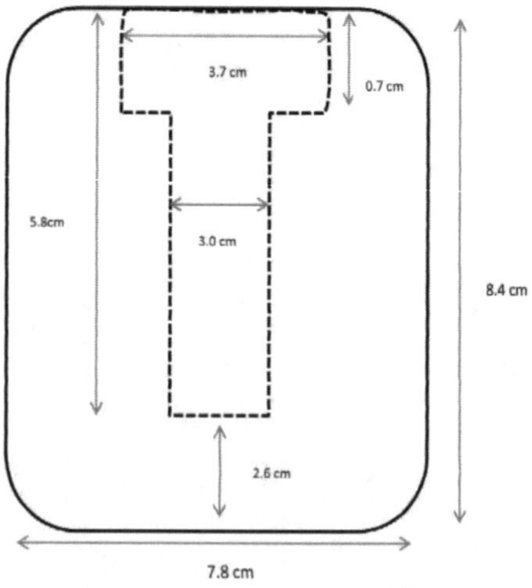

(c) Lead pot (cap and body) dimensions

Figure 3. X-ray images of the lead pot container

In Figure 4, the x-ray image of the nuclear moisture density gauge shows a clear internal view of the equipment; the top side are the electronics and display/keypad, the rear side are the mechanical handle assembly and source housing. Connecting cables to the three detectors are seen in figure 4(a). The lead cup and source rod assembly are clearly seen as well. The x-rays easily passed through the top shell, which is made of polycarbonate casing. The degree of contrast in this image is determined by the disparities in both the density and thickness of the body's structures. The more significant the difference in either density or thickness between neighbouring structures, the greater the contrast between them within the image. Unexpected density or thickness variations in the internal mechanical structures may indicate anomalies or tampering, however, none are detected in this equipment set.

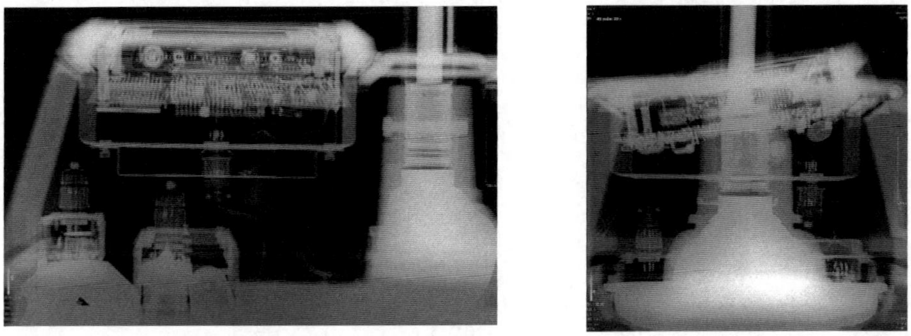

(a) Side view (b) Back view

Figure 4. X-ray images of the nuclear moisture density gauge

4. Conclusion

This study successfully employed a multi-technique laboratory approach to characterise radioactive source casings for nuclear forensic applications. By integrating visual inspection, physical characterisation, SEM-EDX, and X-ray radiography, it provided detailed findings into the casings' physical and chemical properties, offering information on material composition, structural characteristics, and environmental interactions. Findings such as wear, corrosion, and surface oxidation highlighted forensic signatures essential for tracking and regulating radioactive materials. This approach demonstrated the value of detailed casing analysis in strengthening nuclear security and supporting forensic investigations, laying a foundation for future advancements in radioactive source tracking.

5. References

[1] New Straits Times 2018, August 11. Exclusive: Radioactive device missing. New Straits Times. https://www.nst.com.my/news/nation/2018/08/403235/exclusive-radioactive-device-missing

[2] Atom Malaysia. Radiological incident/accident statistics. AELB. Retrieved March 15, 2024 from https://www.aelb.gov.my/v2/index.php/en/radiological-incident-accident-statistics/

[3] Keegan E, Kristo M J, Toole K, Kips R and Young E 2015. Nuclear Forensics: Scientific analysis supporting law enforcement and nuclear security investigations. *Analytical Chemistry* **88**:1496-1505.

[4] Goldberg M, Finck M , Chamberline, D and Carney, K 2011 *Forensic signature of radiological sources.* 52nd Annual Meeting of the Institute of Nuclear Material Management, 2011 (INMM52), Palm Desert, California, USA, 17-21 July 2011

[5] IAEA, 2015 *Nuclear forensic in support investigation* IAEA Nuclear Security Series No. 2-G (Rev. 1). Vienna. International Atomic Energy Agency.

[6]InstroTek Inc. CPN MC3 Elite Moisture/Density Gauge Operating Manual. https://www.instrotek.com/

[7] Sentinel Asia DR FXG-DR1012. www.sentinelndt.com

[8] Golden Engineering XRS4 18V X-ray Generator Operator's Manual. 2017.

[9] IAEA, 2012 *Review of Sealed Source Designs and Manufacturing Techniques Affecting Disused Source Management* Vienna IAEA -TECDOC-1690 International Atomic Energy Agency.

[10] Fontana, M G 1987 *Corrosion Engineering* (3rd ed.). McGraw-Hill.

[11] IAEA, 2002 *Effects of radiation and environmental factors on the durability of materials in spent fuel storage and disposal* Vienna IAEA -TECDOC-1316 International Atomic Energy Agency.

[12] Kondo M, Ishii M, Norimatsu T, Muroga T. 2017 Experimental study on corrosion and precipitation in non-isothermal Pb-17Li system for development of liquid breeder blanket of fusion reactor. *Journal of Physics: Conference Series*. **877**:012001.

[13] Lyon, S B 2010. *Corrosion of lead and its alloys*. In B. R. (Ed.), Corrosion (pp. 245-268). Elsevier. DOI: 10.1016/B978-044452787-5.00098-6.

Acknowledgments
The authors would like to thank International Atomic Energy Agency (IAEA) for the financial support from CRP-IAEA Research Contract No: 26731 and also would like to express appreciations to the management of Malaysian Nuclear Agency for their continuous assistance, advices and supports on this project.

IOP Conf. Series: Materials Science and Engineering **1326** (2025) 012011 doi:10.1088/1757-899X/1326/1/012011

Depth dose measurement by using Al_2O_3 OSL dosimeters in medium with the presence of air cavity and density inhomogeneity on high energy electrons

Mohd Fahmi Mohd Yusof[1,2], Farahamiza Othman[1], Nur Syazwany Malek[1], Arifah Nazirah Abdullah[1], Ahmad Bazlie Abdul Kadir[2], Norriza Mohd Isa[3]

[1]School of Health Sciences, Universiti Sains Malaysia, 16150 Kota Bharu, Kelantan
[2]Faculty of Health & Life Science, Management & Science University, 40100 Shah Alam, Selangor
[3]Medical Physics and Secondary Standard Dosimetry Laboratory, Malaysian Nuclear Agency, 43000 Selangor, Malaysia

mohdfahmi_mohdyusof@msu.edu.my

Abstract. Air cavities and tissue density inhomogeneity significantly affects the distribution of radiation doses, potentially resulting in adverse consequences such as cancer recurrence. This research aims to assess the accuracy of Al2O3 optically stimulated luminescence (OSL) dosimeters in measuring doses within varying thicknesses of air cavities and tissue inhomogeneity. An expanded polystyrene (EPS) was used to simulate the air cavity within the solid water phantoms. The low and high density mediums are simulated by the cork simulating the lungs (5 cm) and bone phantoms (3 cm). The percentage depth-dose (PDD) curves at 9 and 15 MeV electrons were obtained in both presence of air cavity and density inhomogeneity and compared to that in the EBT3 radiochromic film dosimeters and treatment planning system (TPS). The results indicated that the presence of an air cavity and tissue inhomogeneity affected the depth dose measured in OSL dosimeters, EBT3 films and TPS. OSLD and TPS showed good agreement at the centre of the cavity, which is within ±5% but could not estimate scattered radiation to the distal and proximal surfaces of the air cavity. The obtained p-values showed no significant differences of dose measured in OSL dosimeters compared to those in EBT3 films and TPS. The overall results indicated the suitability of OSL dosimeters as indirect dosimeters for the measurements of depth dose in the presence of air cavity and tissue density inhomogeneity.

1. Introduction

Electron beams offer advantages over photon beams due to their higher linear energy transfer (LET) and reduced penetration power, making them suitable for treating superficial or semi-deep-seated tumors near the skin's surface, such as radiation boost to the breast tumor bed (Jong et al., 2018). The combination of high surface dose, coupled with a rapid dose reduction beyond the 80% dose range leads to an overall improvement in the likelihood of tumor control and a decrease in normal tissue complications for superficial tumors (Mutsakanyi & du Plessis, 2021). The effectiveness of the treatment depends on the accuracy of the treatment planning process, but the precision of dose calculation tends to be compromised when dealing with an inhomogeneous density medium (Parenica et al., 2019).

The human body exhibits inhomogeneity due to its composition of diverse tissues and cavities, each having distinct densities, atomic numbers, scattering, and absorption properties (Verma et al., 2019). A prime example is the chest area, which comprises high-density bones, soft tissue, low-density lung tissue, and air cavities. These variations would affect the distribution of doses within the irradiated volume and must be considered during the development of the treatment plan due to the backscattering process (Azizi et al., 2019). High-density tissue such as bone attenuates electron beams, leading to a reduction in dose beyond the bone and an increased dose in front of the bone (Parwaie et al., 2022). In contrast, there is less energy absorption by low-density tissue such as the lung thus increasing the absorbed dose beyond it (Zabihzadeh et al., 2020). Since dose calculations can be imprecise when tissue inhomogeneity is disregarded, measuring dose perturbation within, beyond, in front of, and laterally around the inhomogeneity is crucial to ensure accurate dose delivery and effective treatment.

Dose perturbation can be determined by measuring the percentage depth dose (PDD) because the information about absorbed dose distribution within a patient is commonly characterized by PDD and beam dose profiles or off-axis ratios (Bencheikh et al., 2020). PDD is a critical parameter that helps evaluate a medium's attenuation properties and a dosimeter's detection characteristics. PDD is often measured in a homogeneous medium such as a water phantom by using ionization chamber and radiochromic film dosimeters (Robinson et al., 2020)

2. Methodology

2.1. Depth dose measurements in inhomogeneous density medium

The OSL dosimeters type nanodot™ used in the study were calibrated at high energy photons by using ^{137}Ce gamma energies based by the Malaysian Nuclear Agency (Nuclear Malaysia) as the secondary standard dosimetry laboratory (SSDL). The inhomogeneous density medium was prepared with the simulations of the presence of lung and bone tissues within the solid water phantoms. The lung was simulated by using cork with average density of 0.24 g/cm^3 while the compact bone phantom with density of 1.85 g/cm^3 was used to simulate the bones. Both lung and bone phantoms were sandwiched between the solid water phantoms as shown in Figure 1. The phantoms were irradiated to 9 MeV electrons at 10 × 10 cm field size and 100 cm source to surface distance (SSD). Prescribed dose of 300 cGy given by the monitor unit (MU). A number of depth dose measurements were made in several locations within the mediums including the interface regions (proximal and distal surfaces) and the interface between lung and bone by using OSL dosimeters. A bolus was applied on the surface the OSL dosimeters were placed to prevent the air gap between the phantoms.

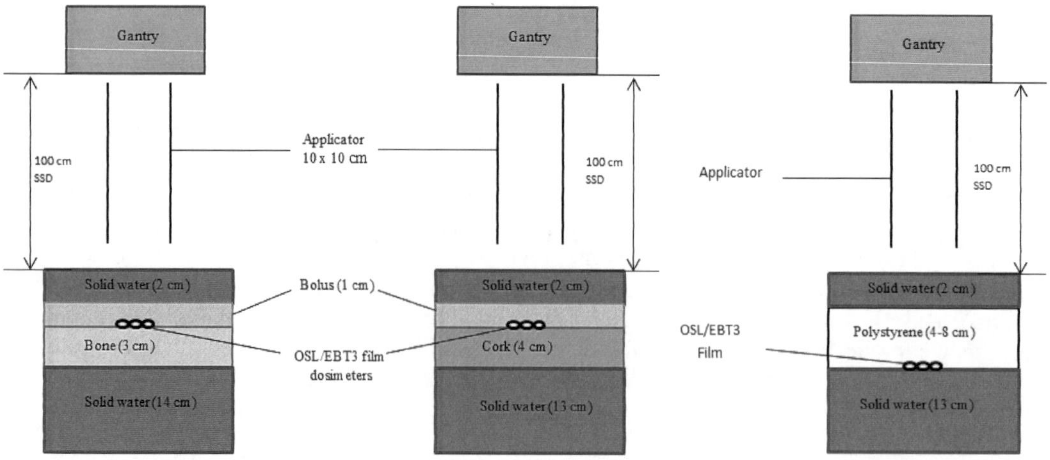

Figure 1. The experimental set for measurement of PDD in inhomogeneous density phantoms of (a) solid water-bone, (b) solid water-lung and (c) presence of air cavity.

2.2. Depth dose measurement in the presence of air cavity

Polystyrene with the density of 0.01 g/cm^3 was used to simulate the air cavity (Song et al., 2020). The polystyrene was sandwiched between the solid water phantoms with various thicknesses between 3and 8 cm as shown in Figure 1(c). The air cavities were simulated at thicknesses of 3, 5 and 8 cm. The phantoms were irradiated to 6 MV photons with 300 cGy prescribed dose at 10 × 10 cm field size and 100 cm source to surface distance (SSD) based on the standard calibration condition of the linear accelerator. The depth doses were measured in several locations within the mediums including the interface regions (proximal and distal surfaces) between solid water phantoms and polystyrene as well as within the polystyrene.

The deviation of depth doses obtained in the OSL were compared to that in the EBT3 film dosimeter by measuring the percentage of differences using the equation:

$$percentage\ difference(\%) = \frac{OSL\ reading - EBT3\ reading}{EBT\ reading} \times 100\% \quad (1)$$

3. Results and discussion

The PDD in OSL dosimeters and EBT3 film dosimeter in the presence of bone and lung phantoms as higher and lower density materials compared to solid water phantoms are illustrated in Figure 2. The PDD of OSL in homogeneous density medium of solid water phantoms showed excellent agreement to EBT 3 film dosimeter (Pimpinella et al., 2019; Singh et al., 2019). The presence of higher and lower density tissue simulated by the bone and lung phantoms significantly altered the PDD curve in the electron beams compared to the homogeneous density solid water phantoms. The presence of high density bone phantoms caused rapid PDD fall-off observed in OSL dosimeters and EBT 3 film dosimeters as shown in Figure 2 (b). The detail of PDD in OSL dosimeters and EBT3 film dosimeters are shown in Table 1. The PDD dose in OSL showed excellent agreement to EBT3 film dosimeter until the depth of 4 cm and within the bone phantoms before the percentage of differences significantly increased (Yusof et al., 2022). Figure 2(b) showed the PDD of OSL dosimeters in solid water-lung phantoms in comparison to the EBT3 film dosimeter. The presence of lower density material compared to solid water has lowered the attenuation of electrons with the medium, reducing the PDD fall-off beyond the maximum dose, d_{max} and extending the practical range of the electrons (Zabihzadeh et al., 2020). The PDD fall-off however was more significant recorded by the OSL dosimeter compared to that in EBT3 film. The detail of PDD in OSL and EBT3 film dosimeter is shown in Table 2.

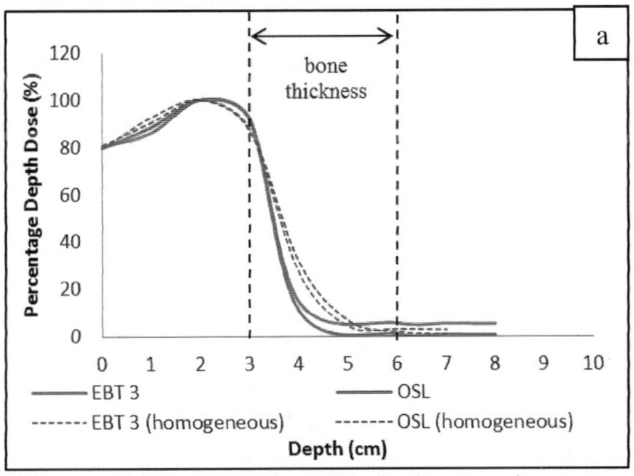

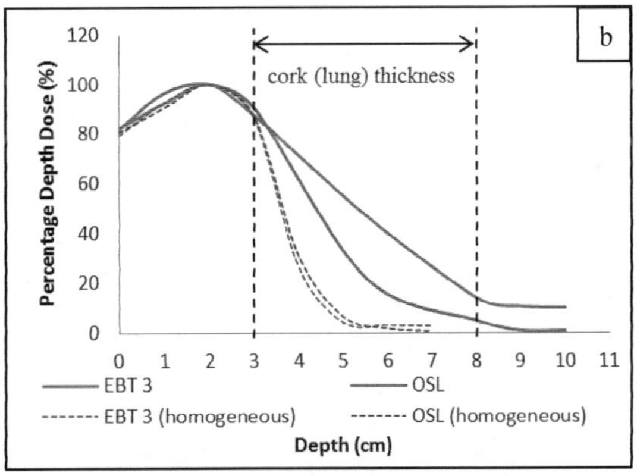

Figure 2. The PDD in OSL dosimeters in comparison to EBT 3 film dosimeter in (a) homogeneous solid water (b) solid water-bone and (c) solid water-lung phantoms.

Table 1. The depth dose measured in the OSL dosimeters and EBT3 film dosimeters at similar depths in inhomogeneous densities of solid water-bone phantoms.

Depth (cm)	Description	Percentage depth dose (%) OSL	EBT3 Film	Percentage Difference (%)
0		79.7	80.08	0.47
1	Solid water phantoms	88.53	86.25	2.64
2		100	100	0.00
3		91.34	91.78	0.48
4	Bone phantoms	10.42	14.25	26.88
6		0.84	5.56	84.89
7		0.82	5.41	84.84
8	Solid water phantoms	0.76	5.3	85.66

Table 2. The depth dose measured in the OSL dosimeters and EBT3 film dosimeters at similar depths in inhomogeneous densities of solid water-lung phantoms.

Depth (cm)	Description	Percentage depth dose (%) OSL	EBT3 Film	Percentage Difference (%)
0	Solid water phantoms	81.93	82.41	0.58
1		96.54	92.31	4.58
2		100	100	0.00
3		90.78	87.67	3.55
6	Lung phantoms	22.27	47.45	53.07
8		5.12	14.22	63.99
9		1.05	10.69	90.18
10	Solid water phantoms	0.95	10.25	90.73

Table 3 showed the electron parameters in solid water-bone and solid water-lung phantoms measured in OSL and EBT3 film dosimeters in comparison to ionization chamber in water. The electron parameters were shifted to closer to the surface in solid water-bone phantom and shifted deeper from the surface in solid water-lung phantoms. The results also showed good agreement of readings between OSL and EBT3 film dosimeters.

Table 3. The electron parameters measured in OSL dosimeters in comparison to EBT3 film dosimeter in solid water-bone and solid water-lung phantoms

Electron parameter	Depth (cm) Water (ionisation chamber)	Solid water – bone OSL	Solid water – bone EBT3	Solid water - lung OSL	Solid water - lung EBT3
R_{90}	2.7	3.0	3.0	3.0	3.0
R_{80}	3.0	3.2	3.2	3.6	3.4
R_{50}	3.5	3.4	3.4	5.0	5.2
R_p	4.3	3.8	3.9	7.6	8.0

Figure 3 illustrated the PDD curve of OSL dosimeters in comparison to EBT3 film in the presence of air cavities of 4, 6 and 8 cm thicknesses in 9 MeV electrons. The presence of air cavities produced plateau-shaped PDD curves observed in all air cavity thicknesses (Sabater et al., 2014). The PDD in OSL dosimeters showed small discrepancies to EBT3 film in the initial thickness of air cavities and increased when the depth of air cavities increased (Moreno et al., 2013). The percentage of discrepancies between OSL and EBT 3 film dosimeters however reduced in thicker air cavity with the lowest was observed in 8 cm air cavity thickness. The presence of air cavities was also observed to extend the practical range, R_p of high energy electrons.

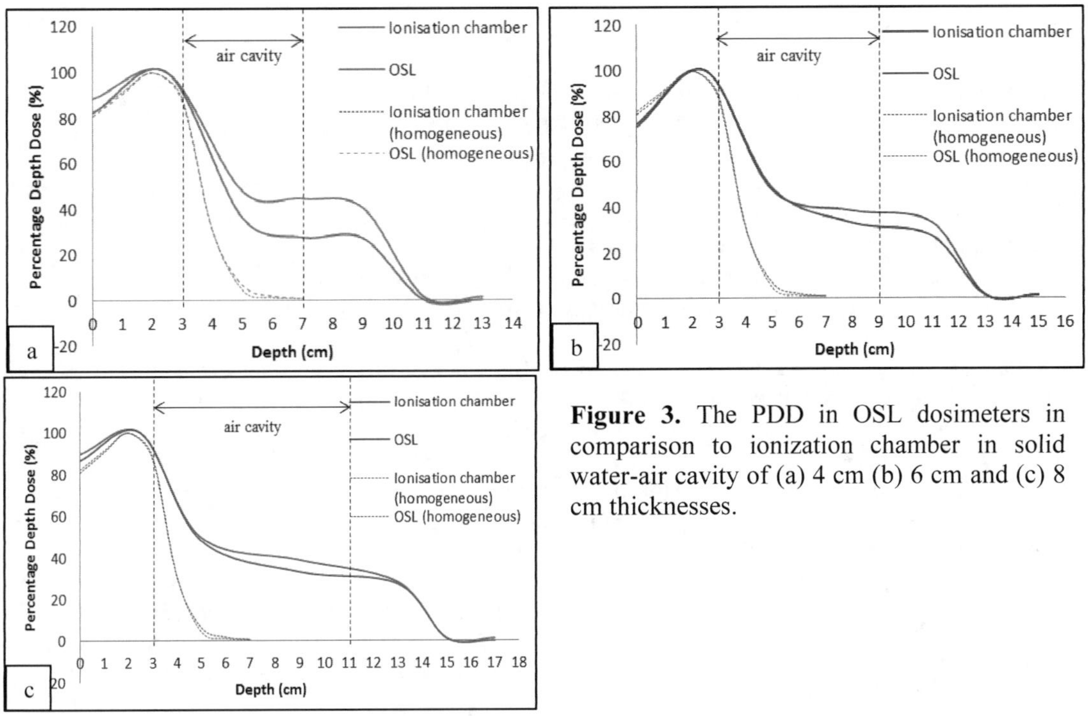

Figure 3. The PDD in OSL dosimeters in comparison to ionization chamber in solid water-air cavity of (a) 4 cm (b) 6 cm and (c) 8 cm thicknesses.

The PDD in the middle of air cavity however were unable to be measured by using both OSL dosimeter and EBT film dosimeter duue to the inability to place the OSL and EBT film dosimeters in the mid-air positions. The reading in the area is useful to understand the pattern of dose pertubation and variations resulted from the existence of the air cavities.

4. Conclusion

The results showed good agreement of dosimetry readings between OSL dosimeters and EBT3 film dosimeter in the inhomogeneous density mediums of solid water-bone and solid water-lung phantoms. The OSL dosimeters also showed good agreement of dosimetry in the presence of air cavities in solid water phantoms and the percentage of discrepancies between OSL dosimeters and EBT3 film dosimeter decreased at thicker air cavities. The overall results indicated the suitability of OSL dosimeters for dosimetry works involving mediums of density inhomogeneity and the presence of air cavities in high energy electrons. The use of several other common dosimeters such as thermoluminescence dosimeter (TLD) and the use of simulation software such as Monte Carlo simulation would provide better understanding of the effect on density inhomogeneity and presence of air cavities in high energy electron as well as the reliability of OSL dosimeter as indirect method of dose measurements.

5. References

[1] Jong W L, Ung N M, Tiong A H, Rosenfeld A B and Wong J H D 2018 Characterisation of a MOSFET-based detector for dose measurement under megavoltage electron beam radiotherapy. *Radiat. Phys. Chem.*, **144**; 76-84.

[2] Mustakanyi S and du Plessis 2020 Characterization and dosimetry of photon multileaf collimated electron beams using Gafchromic film measurements. *Radiat. Phys. Chem.*, **182**; 109266.

[3] Parenica H M, Mayroidis P, Jones W, Swanson G, Papanikolaou N and Stathakis S 2019 VMAT Optimization and Dose calculation in the presence of metallic hip prostheses. *Technol. Cancer Res. Treat.,* **18**; 1533033819892255.
[4] Verma R T, Painuly K N, Mishra P S, Yoganathan S A, Singh N, Bhatt M L B and Jamal N 2019 Evaluation of lung density and its dosimetric impact on lung cancer radiotherapy: A simulation study. *J. Biomed. Phys. Eng.,* **9(1)**; 17-28.
[5] Azizi M, Mowlavi A, Ghorbani A, Knuap C and Behmadi M 2019 A Monte Carlo study on dose perturbation due to dental restorations in a 15 MV photon beam. *J. Cancer Res. Ther.,* **15(3)**; 491–497.
[6] Parwaie W, Geraily G, Mehri-Kakavand G, Babaloui S, Rezvani S and Pursamimi M 2022 Dosimetry of small photon fields in the presence of bone heterogeneity using MAGIC polymer gel, Gafchromic film, and Monte Carlo simulation. *Rep. Pract. Oncol. Radiother.,* **27(2)**; 226–234.
[7] Zabihzadeh M, Ghahremani Z, Hoseini S M, Shahbazian H and Hoseini G M 2020 Effect of lung inhomogeneity on dose distribution during radiotherapy of patient with lung cancer. *Int. J. Radiat. Res.*, **18(3)**; 579–586.
[8] Bencheikh M, Maghnouj A and Tajmouati J 2020 Dosimetry quality control based on percent depth dose rate variation for checking beam quality in radiotherapy. *RPOR*, **25(4)**; 484–488.
[9] Robinson S M, Esplen N, Wells D and Bazalova-Carter M 2020 Monte Carlo simulations of EBT3 film dose deposition for percentage depth dose (PDD) curve evaluation. *J. Appl. Clin. Med. Phys.,* **21(12)**; 314–324.
[10] Song Y K, Hong S H, Eo S, Han M G and Shim W J 2020 Rapid Production of Micro- and Nanoplastics by Fragmentation of Expanded Polystyrene Exposed to Sunlight. *Environ. Sci. Technol.,* **54(18)**; 11191–11200.
[11] Pimpinella M, Andreoli S, De Angelis C, Della Monaca S, D'Arienzo M. and Menegotti L. 2019 Output factor measurement in high dose-per-pulse IORT electron beams. *Phys. Med.,* **61**; 94–102.
[12] Singh S, Semwal M and Bhatt C 2019 Estimation of backscatter from internal shielding in electron beam radiotherapy using Monte Carlo simulations (EGSnrc) and Gafchromic film measurements. *J. Med. Phys.*, **44(4)**; 239–245.
[13] Yusof M F M, Shariff N A, Fauzi N S M, Kadir A B A, and Isa N M 2022 Small beam dosimetry by using Al_2O_3 optically stimulated luminescent (OSL) dosimeters at high energy photons and electrons. *IOP Conf. Series: Mater. Sci. Eng.,* **1231(1)**; 012018.
[14] Sabater S, Berenguer R, Honrubia-Gomez P, Rivera M, Nunez A, Jimenez-Jimenez E, Martos A and Ramirez-Castillo C 2014 How air influences radiation dose deposition in multiwell culture plates: a Monte Carlo simulation of radiation geometry. *J.R.R..;* **2014** (1-6).
[15] Zarza-Moreno M, Carreira P, Madureira L, Miras del Rio H, Salguearo F J, Leal A, Teixeira N, Jesus A P and Mora G (2014) Dosimetric effect by shallow air cavities in high energy electron beams, *Phys. Med.,* **30(2)**; 234-241.

6. Acknowledgement

The authors would like to acknowledge the financial support from the USM Research Universiti (RU-I) grant scheme no 1001.PPSK.8012381. The authors also would like to acknowledge the Nuclear Medicine, Radiotherapy and Oncology Department, Hospital Universiti Sains Malaysia (HUSM) for the use of the linear accelerator software in this study.

IOP Conf. Series: Materials Science and Engineering **1326** (2025) 012012 doi:10.1088/1757-899X/1326/1/012012

Assessments of natural radionuclides concentration and radiological hazard indices in Universiti Teknologi Malaysia (UTM)

Ko Yu Yun[1], Muhammad Arif Bin Sazali[1,a)]

[1]Faculty of Chemical and Energy Engineering, UTM 81310 Skudai, Johor, Malaysia

[a)]muhammadarifsazali@utm.my

Abstract. Terrestrial gamma radiation arises from the decay of naturally occurring radionuclides, known as primordial radionuclides. The radiological hazard associated with these radionuclides can be quantified using radiological hazard indices. This study aims to (1) determine the average Terrestrial Gamma Radiation Dose (TGRD) rate at Universiti Teknologi Malaysia (UTM), (2) assess the average concentrations of radionuclides ^{238}U, ^{232}Th, and ^{40}K in UTM soil, and (3) evaluate the corresponding radiological hazard indices. The mean TGRD rate, measured at 30 locations across UTM, was 100.11 ± 16.22 nGy/h, which exceeds the national average of 92 nGy/h and the global average of 59 nGy/h. Gamma spectra of soil samples from five locations with the highest TGRD readings were analysed using a high-purity germanium (HPGe) detector. The mean activity concentrations of ^{238}U, ^{232}Th, and ^{40}K were 71.29 ± 33.11 Bq/kg, 74.09 ± 27.72 Bq/kg, and 343.03 ± 226.52 Bq/kg, respectively. The calculated radiological hazard indices—annual effective dose equivalent, radium equivalent activity, external hazard index, and internal hazard index—were all below international safety limits, indicating that the radiation hazard levels in UTM are within safe bounds.

1. Introduction

Humans are inevitably exposed to ionising radiation emitted from various natural sources. Terrestrial gamma radiation refers to the natural ionising radiation emitted during the decay of natural radionuclides, also known as primordial radionuclides. Uranium-238 (^{238}U), thorium-232 (^{232}Th), and potassium-40 (^{40}K) are examples of such radionuclides found in soil, water, and other environmental sources. At sea level, terrestrial gamma radiation constitutes the major source of external radiation exposure compared to cosmic gamma radiation [1]. As the terrestrial radiation dose is dependent on the concentration of these primordial radionuclides in the geological formation, consequently, studies on natural radiation predominantly focus on determining the Terrestrial Background Radiation Dose (TGRD) rates or the radioactivity of associated sources [2–7]. Such investigations are beneficial in assessing the public TGRD rates and conducting epidemiological studies, as well as creating a database for tracking the change in environmental radioactivity [8].

In addition to emitting gamma radiation, the decay series of ^{238}U and ^{232}Th generates alpha-emitting radon gas, which is the primary contributor to internal exposure to natural radiation. While alpha radiation from external exposure is blocked by the skin, gamma radiation can penetrate the body and cause DNA damage. Both gamma and alpha radiation are harmful ionising radiations that can potentially damage or mutate healthy cells in the human body. To evaluate the potential hazards associated with the decay of these radionuclides, radiological hazard indices are commonly employed in radioactivity studies. The activity concentrations of these radionuclides are used to calculate indices such as the annual effective dose rate and internal hazard index, which could estimate the potential risk posed by radon and its decay products to the respiratory system [3].

Universiti Teknologi Malaysia (UTM) in Skudai, Johor, spans an area of 12.22 km^2 and hosts over 10,000 full-time undergraduate students. Despite the large population of students and staff residing on campus daily, there is a lack of comprehensive studies on natural radiation exposure in this environment. Therefore, this study aims to: (1) determine the mean TGRD rate at UTM Skudai by measuring dose rates at 30 selected sites using the SEI Inspector USB Handheld Radiation Alert ® Detector, (2) determine the mean activity concentrations of ^{238}U, ^{232}Th, and ^{40}K through gamma spectroscopy analysis of soil samples collected from five sites with the highest TGRD readings, and (3) assess radiological hazard indices in the UTM Skudai area using the mean activity concentrations of the aforementioned radionuclides. The indices evaluated include the annual effective dose equivalent (*AEDE*), radium equivalent activity (*REA*), external hazard index (H_{ext}), and internal hazard index (H_{int}).

This study establishes a foundation for future relevant research in UTM and Johor by contributing to the database on natural radiation exposure. By identifying the TGRD rates and associated hazard indices, this research enhances the among UTM staff and students about the radiological risk levels in their environment. This research could also potentially encourage other universities to assess and address radiation exposure risks on their campuses.

2. Methodology

2.1. Measurement of TGRD rates in UTM

Figure 1 illustrates the 30 measuring sites across the UTM Skudai campus that were selected to determine the mean TGRD rate. The sites were chosen based on the following criteria: (1) the ground surface was unpaved, (2) no man-made buildings or structures were located within a 5-meter radius, and (3) the soil at the site had not been transported from other locations [9]. TGRD measurements at these sites were conducted using the SEI Inspector USB Handheld Radiation Alert® Detector with gamma detection sensitivity down to 10 keV and an accuracy of ± 0.001 mR/hr. To minimise the influence of environmental influences such as radon progeny washout from the air, all the measurements were conducted after at least 12 hours of sunny weather [1]. During the measurement, the detector was positioned 1 meter above the ground. Five readings were recorded at 1-minute intervals at each site. Each measuring point was spaced at least 1 meter apart from the others to ensure spatial variation.

2.2. Determination of activity concentration of ^{238}U, ^{232}Th and ^{40}K

Based on the average TGRD rates from the 30 locations, five sites with the highest TGRD rates were selected for soil sample collection. This selection strategy aimed to identify areas with potentially higher concentrations of primordial radionuclides, enabling a more focused assessment of the potential radiological hazards in UTM. At each site, the top 3–5 cm layer of surface soil, which might be contaminated by radionuclide fallout from rain, was removed before collecting approximately 1 kg of soil. The samples were first dried in an oven for 24 hours at 120 ºC. Once dried, the soil samples were crushed and ground into fine powder, then sieved using a 250 μm sieve. The sieved soil was weighed and sealed in Marinelli beakers, where it was left to sit for at least a month to attain secular equilibrium between the long half-life parent radionuclides and their daughter radionuclides [10].

Figure 1. Map of UTM Skudai campus showing the selected locations for TGRD rate measurements. Green markers indicate all the initially selected sites for TGRD measurements, while white markers represent the specific locations where both TGRD measurements and soil samples were collected after determining the mean TGRD rates.

After a month, gamma spectroscopy analysis was conducted on the soil samples using the ORTEC coaxial HPGe detector. Prior to analysis, the HPGe detector underwent energy and efficiency calibration. Each sample was then counted individually, with a counting time of 24 hours for each. Using the gamma spectra obtained, the activity concentrations of ^{238}U, ^{232}Th, and ^{40}K were calculated using equation (1). In this equation, A_i represents the activity concentration of radionuclide i in Bq/kg, C_{Ei} is the number of counts at energy E for radionuclide i, C_{eff} is the detection efficiency at energy E, γ is the gamma emission probability (%) for radionuclide i at energy E, m is the mass of the sample in kg, and t is the counting time in seconds. The gamma energy peaks and emission probability used for activity concentration calculations are listed in Table 1 below [11]. Both activity concentrations of ^{238}U and ^{232}Th are calculated from the mean activity concentration of their gamma-emitting daughter radionuclides.

$$A_i = \frac{C_{Ei}}{C_{eff} \times \gamma \times m \times t} \tag{1}$$

Table 1. Gamma energy peaks and emission probability for ^{238}U, ^{232}Th, and ^{40}K activity concentration calculation.

Primordial Radionuclides	Daughter Radionuclides	Gamma Peak (keV)	Emission Probability (%)
^{238}U	^{214}Bi	609.2	48.0
	^{214}Pb	351.96	39.3
^{232}Th	^{228}Ac	911.2	25.80
	^{208}Tl	583.19	84.50
^{40}K	-	1461.0	10.72

2.3. Radiological hazard indices

Annual effective dose equivalent (*AEDE*) is the calculation of the annual effective dose that will be received by the public with the consideration of an outdoor occupancy factor of 0.2. This factor assumes that most people spend 20 % of their time outdoors exposed to terrestrial gamma radiation directly [12]. According to the International Commission on Radiological Protection (ICRP), the recommended dose limit for the public is 1 mSv/y [13]. Therefore, 1 mSv/y serves as the safe limit for the *AEDE* value. Equation (2) and equation (3) will be used to calculate the *AEDE*, where D represents the absorbed dose rate due to gamma radiations in the air at 1 m above the ground surface, assuming a uniform distribution of primordial radionuclides [6].

$$D = 0.462A_{U-2} \quad + 0.604A_{Th-23} \quad + 0.0417A_{K-4} \tag{2}$$

$$AEDE = D\left(\frac{nGy}{h}\right) \times 0.2 \times 0.7\left(\frac{Sv}{Gy}\right) \times 24hours \times 365.25days \times 10^{-6} \tag{3}$$

Radium equivalent activity (*REA*) defines the radiation exposure to humans from primordial radionuclides by considering the exposure from a uniform distribution of ^{238}U, ^{232}Th and ^{40}K in the soil as a single quantity [8, 14–15]. Equation (4) shown below was used for the *REA* calculation, with the recommended maximum value of *REA* being 370 Bq/kg [15]. This limit ensures that the dose contribution by these primordial radionuclides will not exceed 1.5 mSv/y [15].

$$REA = A_{U-238} + 1.43A_{Th-232} + 0.077A_{K-40} \tag{4}$$

External hazard index (H_{ext}) is the measure of the indoor radiation dose rate due to external exposure to gamma radiation from primordial radionuclides in the soil. This index value should be less than 1 to ensure that the radiation hazard remains within a safe level [16]. Equation (5) was used for the calculation of H_{ext}.

$$H_{ext} = \frac{A_{U-238}}{370Bq/kg} + \frac{A_{Th-232}}{259Bq/kg} + \frac{A_{K-40}}{4810Bq/kg} \quad (5)$$

Internal hazard index (H_{int}) is used to evaluate the potential risk posed by radon and its decay products to the respiratory system. For radiological safety, the internal hazard index should be less than 1 as well [3, 6]. Equation (6) was used for the calculation of H_{int}.

$$H_{int} = \frac{A_{U-238}}{185Bq/kg} + \frac{A_{Th-23}}{259Bq/kg} + \frac{A_{K-40}}{4810Bq/kg} \quad (6)$$

3. Result and discussion

3.1. TGRD rates in UTM

The results of the TGRD rate measurements are presented in Table 2. The highest recorded TGRD rate was at Stadium Azman Hashim, with a value of 170.32 ± 16.22 nGy/hr, while the lowest was at KLG Campus Residence, with a reading of 57.35 ± 11.66 nGy/hr. On average, the TGRD rate at UTM is 100.11 ± 16.22 nGy/hr. This average is lower than the TGRD rates observed in Johor and its districts, as detailed in Table 3. However, it exceeds both the Malaysian average of 92 nGy/hr and the global average of 59 nGy/hr.

Even though the mean TGRD rate in UTM is slightly higher than the global average, this is still considered low compared to the dose rates in High Background Natural Radiation (HBNR) areas. HBNR areas such as YangJiang in China and Kerala in India have reported an average annual effective dose exceeding 5 mSv/y [1, 17]. With a conversion factor of 0.7, the annual effective dose rate in UTM is 0.61 mSv/y. Based on the criteria of classification proposed by Sohrabi. M., areas with potential public exposure less than 5 mSv/y are considered as Low or Normal Level Natural Radiation Areas (LLNRA). Therefore, the TGRD rates in UTM can still considered at a normal level, and no remedial action is required [18].

TGRD rates are significantly influenced by the underlying geology. For example, a study done by Ramli et al. has shown that Haplic Acrisol-Ferric Acrisol soils, derived from granite parent material, exhibit higher TGRD rates [7]. Areas with extensive granitic intrusions in their geological background also tend to have higher TGRD rates [5, 7, 19–20]. This correlation is attributed to the higher concentrations of radioactive minerals naturally occurring within igneous rocks like granite [1]. Moreover, other soil types, such as monazite sand deposits, which are rich in thorium, and those containing thorium- and uranium-bearing minerals, can exhibit elevated TGRD rates [1].

As this study did not specifically investigate the geological formations within the UTM campus, the discussion on factors affecting TGRD rates will focus on correlating them with the activity concentrations of primordial radionuclides in the soil, as presented in Section 3.2.

Table 2. TGRD rate for 30 locations in UTM and mean TGRD rate in UTM.

Number	Location	Average Reading (nGy/hr)
1	Stadium Azman Hashim	170.32 ± 13.18
2	L12, Kolej Tun Hussein Onn (KTHO)	145.99 ± 20.75
3	Kompleks Sukan UTM	135.56 ± 20.93
4	MA1, Kolej Tun Dr Ismail (KTDI)	128.61 ± 14.28
5	Institute of Bioproduct Development	125.14 ± 23.48
6	S01, Kolej Tunku Canselor (KTC)	112.97 ± 25.34
7	Arked Cengal	112.97 ± 18.43
8	Academic Management Division	109.49 ± 9.91
9	K32, Kolej Tun Razak (KTR)	106.02 ± 15.55
10	Pusat Kesihatan UTM	106.02 ± 18.84
11	UTM Cricket and Football Field	100.80 ± 15.79
12	B03, Faculty of Architecture and Built Environment (FABU)	99.07 ± 14.54
13	UTM Gate 3	95.59 ± 23.80
14	G01, Kolej Rahman Putra (KRP)	95.59 ± 16.26
15	S43, Kolej Tunku Canselor (KTC)	93.85 ± 16.72
16	K01, KTR	93.85 ± 12.89
17	Masjid Sultan Ismail	93.85 ± 7.27
18	P02, Fakulti Kejuruteraan Elektrik (FKE)	92.11 ± 20.01
19	H21, Kolej Tun Fatimah (KTF)	90.38 ± 18.02
20	Pejabat Kor Suksi	88.64 ± 16.72
21	N17, Fakulti Kejuruteraan Kimia dan Tenaga (FKT)	88.64 ± 9.52
22	Bamboo Avenue	86.90 ± 16.26
23	P18, Advanced Microwave and Antenna Laboratory	85.16 ± 14.28
24	Perpustakaan Raja Zarith Sofea (PRZS)	85.16 ± 18.84
25	Padang Bola UTM	85.16 ± 15.55
26	T06, Faculty of Geoinformation and Real Estate	83.42 ± 13.18
27	Kolej Datin Seri Endon (KDSE)	81.69 ± 15.79
28	S36, KTC	81.69 ± 14.54
29	XB1, Kolej Dato Onn Jaafar (KDOJ)	71.26 ± 11.33
30	KLG Campus Residence	57.35 ± 11.66
	Mean TGRD rate	100.11 ± 16.22

Table 3. Comparison of mean TGRD rate in UTM with world and other districts in Johor.

Area/District/State/Country	TGRD rate (nGy/hr)	Reference
UTM	100.11	This study
Johor	163.00	[21]
Johor Bahru District, Johor	135.00	[22]
Mersing District, Johor	140.00	[23]
Segamat District, Johor	276.00	[24]
Kluang District, Johor	144.00	[25]
Malaysia	92.00	[1]
World Average	59.00	[1]

3.2. Activity Concentration of ^{238}U, ^{232}Th and ^{40}K in UTM
The minimum detectable activities (MDAs) for ^{238}U, ^{232}Th and ^{40}K are 0.61 Bq/kg, 13 Bq/kg, and 10.18 Bq/kg, respectively. Analysis of the five soil samples collected revealed that the activity concentration of ^{238}U varied from 30.44 ± 0.58 Bq/kg to 107.79 ± 1.20 Bq/kg, with a mean value of 71.29 ± 33.11 Bq/kg. For ^{232}Th, the activity concentration ranged from 30.79 ± 0.72 Bq/kg to 102.73 ± 1.02 Bq/kg, with a mean concentration of 74.09 ± 27.72 Bq/kg. The mean activity concentration of ^{40}K was 343.03 ± 226.52 Bq/kg, with values ranging between 139.94 ± 3.60 Bq/kg and 655.06 ± 7.06 Bq/kg. Figure 2 illustrates the activity concentrations of the radionuclides across the five sampling locations, along with the mean values calculated for UTM. The data was arranged based on the decreasing trend of the measured average TGRD rates in the chosen sampling locations.

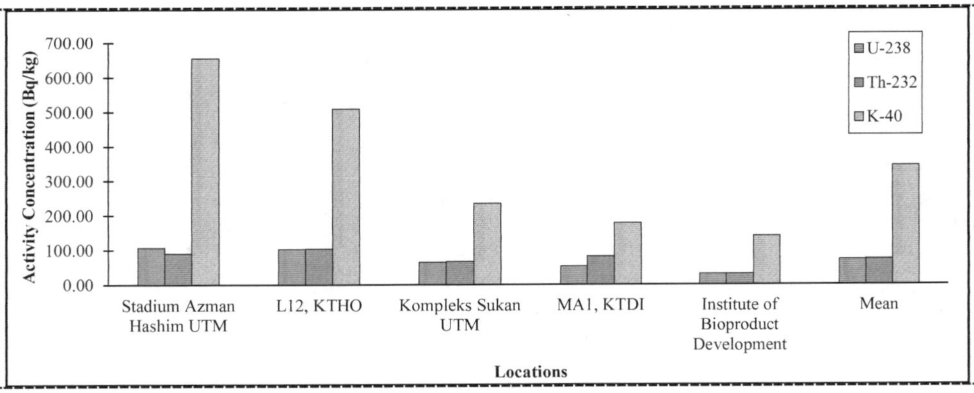

Figure 2. Graph of activity concentration of ^{238}U, ^{232}Th, and ^{40}K at five locations with the highest TGRD rate in UTM.

Comparing the TGRD rates and activity concentration results, areas with higher TGRD rates generally showed higher activity concentrations of ^{238}U and ^{40}K. However, this trend may only apply to areas with high TGRD rates in UTM (≥ 125 nGy/hr). Activity concentration of these radionuclides in areas with lower TGRD rates (< 125 nGy/hr) may not necessarily follow this trend. More sampling locations, including areas with low TGRD rates, should be included to allow a better correlation between the activity concentration and TGRD rates. Extended investigations also needed to correlate the activity concentrations with the geological formations to provide a better understanding of the factors influencing measured TGRD rates in UTM.

3.3. Radiological hazard indices in UTM
According to Table 4, the mean *AEDE* at UTM is 0.11 mSv/year, which is below the ICRP recommended limit of 1.0 mSv/year for public external dose exposure [13]. The mean *REA* at UTM is 203.65 Bq/kg, which is 1.8 times lower than the recommended limit of 370 Bq/kg [14]. The H_{ext} at UTM is 0.55, below the unity threshold, indicating that the soil in these locations does not pose harmful health effects from external gamma radiation exposure. The H_{int} at UTM is 0.74, which is also below the recommended limit of 1.0. Since these indices (*AEDE*, *REA*, H_{ext} and H_{int}) are calculated based on the high TGRD rate areas in UTM as presented in Table 4, the overall radiological hazard level across the campus is likely to be lower than these values, indicating no significant health impacts from the exposure of the terrestrial natural radiation.

Table 4. Radiological hazard indices for five locations with the highest average TGRD rate and the mean radiological hazard indices in UTM.

Location	Radiological Hazard Indices			
	AEDE (mSv/y)	**REA (Bq/kg)**	$\mathbf{H_{ext}}$	$\mathbf{H_{int}}$
Stadium Azman Hashim UTM	0.16	286.76	0.77	1.07
L12, KTHO	0.16	288.32	0.78	1.06
Kompleks Sukan UTM	0.10	175.66	0.47	0.65
MA1, KTDI	0.10	182.26	0.49	0.63
Institute of Bioproduct Development	0.05	85.25	0.23	0.31
Mean	0.11	203.65	0.55	0.74

However, the H_{int} values at Stadium Azman Hashim UTM and L12, KTHO exceed 1.0. Despite these higher values, the activity concentrations of primordial radionuclides at Stadium Azman Hashim and L12, KTHO are not expected to cause significant health effects to students and staff. For the radiological impact of the radionuclide activity concentration to be considered significant, the value of H_{int} would need to be much higher, approximately in the millions [26]. Additionally, further investigations are required to assess the risk of radon exposure from the soil in these areas, as radon emission from primordial radionuclides in the soil to the atmosphere can be affected by several factors. For instance, the soil radon concentrations, radon emanation coefficient, and radon exhalation rate from the soil to the atmosphere will affect the radon exposure risk for the public [27–28]. Therefore, further investigations into these factors at locations with high TGRD rates are advisable to obtain a more accurate estimation of the radon exposure risk compared to relying solely on the value of H_{int}.

Regarding the risk of long-term exposure to terrestrial gamma radiation at UTM, there is no solid evidence to prove the increase of cancer or non-cancer-related diseases at exposure levels below 10 mGy [29]. Given the *AEDE* of 0.11 mSv/y and the measured TGRD rates of 100 nGy/hr, the risk of increased cancer due to this low dose rate is unlikely. Nevertheless, consistent monitoring of TGRD rates is highly encouraged to allow early detection of potential radioactive contamination in the future.

4. Conclusion

This study provides a comprehensive assessment of the TGRD rates and radiological hazard indices at the UTM Skudai campus. The mean TGRD rate across 30 measurement locations was found to be 100.11 ± 16.22 nGy/hr. While this is higher than the Malaysian average of 92 nGy/hr and the global average of 59 nGy/hr, it is lower than the TGRD rates in Johor and its districts. The analysis of soil samples from the five locations with the highest TGRD rates revealed mean activity concentrations of 71.29 ± 33.11 Bq/kg for ^{238}U, 74.09 ± 27.72 Bq/kg for ^{232}Th, and 343.03 ± 226.52 Bq/kg for ^{40}K. The radiological hazard indices computed from these measurements — *AEDE*, *REA*, H_{ext}, and H_{int} — all fall within safe limits. While two areas in UTM (Stadium Azman Hashim UTM and L12, KTHO) have slightly higher H_{int} values, the potential hazard is insignificant and should be further validated with more targeted investigations on radon exposure risk. Overall, the findings suggest that while UTM's TGRD rates are higher than some regional averages, they remain within safe exposure levels for human health. This study underscores the importance of ongoing monitoring and assessment to ensure that radiation levels remain within acceptable bounds and continue to safeguard public health.

5. References

[1] UNSCEAR 2000 *Sources and Effects of Ionizing Radiation, United Nations Scientific Committee on the Effects of Atomic Radiation (UNSCEAR) 2000 Report to the General Assembly, with Scientific Annexes* vol I (New York: United Nations)

[2] Karunakara N, Yashodhara I, Kumara K S, Tripathi R M, Menon S N, Kadam S and Chougaonkar M P 2014 Assessment of ambient gamma dose rate around a prospective uranium mining area of South India–a comparative study of dose by direct methods and soil radioactivity

measurements *Results Phys.* **4** 20–27

[3] Abojassim A A and Rasheed L H 2019 Mapping of terrestrial gamma radiation in soil samples at Baghdad governorate (Karakh side), using GIS technology *Nat. Environ. Pollut. Technol.* **18**(4) 1095–1106.

[4] Agar O, Eke C, Boztosun I and Korkmaz M E 2015 Determination of naturally occurring radionuclides in soil samples of Ayranci, Turkey *J. Phys.: Conf. Ser.* **590** 012042

[5] Gabdo H T, Ramli A T, Saleh M A, Sanusi M S, Garba N N and Aliyu A S 2015 Radiological hazard associated with natural radionuclide concentrations in the northern part of Pahang state Malaysia *Environ. Earth Sci.* **73** 6271–6281.

[6] Al-Trabulsy H A, Khater A E M and Habbani F I 2011 Radioactivity levels and radiological hazard indices at the Saudi coastline of the Gulf of Aqaba *Radiat. Phys. Chem.* **80**(3) 343–348.

[7] Ramli A T, Apriantoro N H and Wagiran H 2009 Assessment of radiation dose rates in the high terrestrial gamma radiation area of Selama District, Perak, Malaysia *Appl. Phys. Res.* **1**(2) 45.

[8] Tufail M, Akhtar N and Waqas M 2006 Measurement of terrestrial radiation for assessment of gamma dose from cultivated and barren saline soils of Faisalabad in Pakistan *Radiat. Meas.* **41**(4) 443–451.

[9] Chikasawa K, Ishii T and Sugiyamab H 2001 Terrestrial gamma radiation in Kochi Prefecture, Japan *J. of Health Sci.* **47**(4) 362–372.

[10] Habib M A, Basuki T, Miyashita S, Bekelesi W, Nakashima S, Phoungthong K and Techato K 2019 Distribution of naturally occurring radionuclides in soil around a coal-based power plant and their potential radiological risk assessment *Radiochim. Acta* **107**(3) 243–259.

[11] Kapanadze K, Magalashvili A and Imnadze P 2019 Distribution of natural radionuclides in the soils and assessment of radiation hazards in the Khrami Late Variscan crystal massif (Georgia) *Heliyon* **5**(3).

[12] UNSCEAR 2010 *Sources and Effects of Ionizing Radiation, United Nations Scientific Committee on the Effects of Atomic Radiation (UNSCEAR) 2008 Report to the General Assembly, with Scientific Annexes A and B* vol I (New York: United Nations)

[13] Cappelen T and Unhjem J F 2009 Chapter 100 - Use of 131-Iodine and the risk of radiation exposure: potential hazards to the patient and other people Eds. Preedy V R, Burrow G N and Watson R *Comprehensive Handbook of Iodine Academic Press* 965–978.

[14] Ravisankar R, Chandramohan J, Chandrasekaran A, Prince Prakash Jebakumar J, Vijayalakshmi I, Vijayagopal P and Venkatraman B 2015 Assessments of radioactivity concentration of natural radionuclides and radiological hazard indices in sediment samples from the East coast of Tamilnadu, India with statistical approach *Mar. Pollut. Bull.* **97**(1-2) 419–430.

[15] Baykara O, Karatepe Ş and Doğru M 2011 Assessments of natural radioactivity and radiological hazards in construction materials used in Elazig, Turkey *Radiat. Meas.* **46**(1) 153–158.

[16] Bavarnegin E, Moghaddam M V and Fathabadi N 2013 Natural radionuclide and radiological assessment of building materials in high background radiation areas of Ramsar, Iran *J. Med. Phys.* **38**(2) 93–97

[17] Hendry J H, Simon S L, Wojcik A, Sohrabi M, Burkart W, Cardis E, Laurier D, Tirmarche M and Hayata I 2009 Human exposure to high natural background radiation: what can it teach us about radiation risks? *J. Radiol. Prot.* **29**(2A) A29.

[18] Sohrabi M 1998 The state-of-the-art on worldwide studies in some environments with elevated naturally occurring radioactive materials (NORM) *Appl. Radiat. Isot.* **49**(3) 169–188.

[19] Abba H T, Hassan W M S W, Saleh M A, Aliyu A S and Ramli A T 2017 Terrestrial gamma radiation dose (TGRD) levels in northern zone of Jos Plateau, Nigeria: Statistical relationship between dose rates and geological formations *Radiat. Phys. Chem.* **140** 167–172.

[20] Abdulkadir M, Garba N N, Nasiru R, Saleh M A, Bello S and Khandaker M U 2023 Statistical analysis of terrestrial gamma radiation dose rates in relation to different geological formations and soil types of Katsina State, Nigeria *Int. J. Environ. Anal. Chem.* **103**(14) 3251–3263.

[21] Ramli A T, Hussein A W M A and Lee M H 2001 Geological influence on terrestrial gamma radiation dose rate in the Malaysian State of Johore *Appl. Radiat. Isot.* **54**(2) 327–333.

[22] Saleh M A, Ramli A T bin Hamzah K, Alajerami Y, Mhareb M H A, Aliyu A S and Hanifah N Z H B A 2015 Natural environmental radioactivity and the corresponding health risk in Johor Bahru District, Johor, Malaysia *J. Radioanal. Nucl. Chem.* **303** 1753–1761.

[23] Saleh M A, Ramli A T, Alajerami Y, Aliyu A S and Bt Basri N A 2013 Radiological study of Mersing District, Johor, Malaysia *Radiat. Phys. Chem.* **85** 107–117.

[24] Saleh M A, Ramli A T, Alajerami Y and Aliyu A S 2013 Assessment of environmental 226Ra, 232Th and 40K concentrations in the region of elevated radiation background in Segamat District, Johor, Malaysia *J. Environ. Radioact.* **124** 130–140.

[25] Aziz Saleh M, Termizi Ramli A, Alajerami Y, Damoom M and Sadiq Aliyu A 2014 Assessment of health hazard due to natural radioactivity in Kluang District, Johor, Malaysia *Isot. Environ. Health Stud.* **50**(1) 103–113.

[26] Rao D 2018 Use of hazard index parameters for assessment of radioactivity in soil: A view for change *Radiat. Prot. Environ.* **41**(2) 59–60.

[27] Vaupotič J, Gregorič A, Kobal I, Žvab P, Kozak K, Mazur J and Grządziel D 2010 Radon concentration in soil gas and radon exhalation rate at the Ravne Fault in NW Slovenia *Nat. Hazards Earth Syst. Sci.* **10**(4) 895–899.

[28] Thu H N P, Van Thang N, Loan T T H and Van Dong N 2019 Natural radioactivity and radon emanation coefficient in the soil of Ninh Son region, Vietnam *Appl. Geochem.* **104** 176–183.

[29] Wojcik A 2022 Reflections on effects of low doses and risk inference based on the UNSCEAR 2021 report on 'biological mechanisms relevant for the inference of cancer risks from low-dose and low-dose-rate radiation' *J. Radiol. Prot.* **42**(2) 023501.

IOP Conf. Series: Materials Science and Engineering **1326** (2025) 012013 doi:10.1088/1757-899X/1326/1/012013

Statistical analysis of disused sealed radioactive sources (DSRS) from radioactive waste inventory in Malaysia

Ahmad Hasnulhadi Che Kamaruddin[1, a)], Nurul Izzah Muhammad Zamri[2]

[1]Waste and Environmental Technology Division, Agensi Nuklear Malaysia, Bangi 43000 Kajang, Selangor, Malaysia

[2]Department of Science Mathematics, Faculty of Science and Technology, National University of Malaysia (UKM), 43600 Bangi, Selangor, Malaysia

[a]hadi@nm.gov.my

Abstract. A Disused Sealed Radioactive Source (DSRS) is a type of sealed radioactive material that is no longer in use or has been removed from service. These sources may have become outdated due to the end of their operational life, advancements in technology, or changes in regulations. If not properly managed and disposed of, DSRS can pose significant risks. Improper storage and handling can lead to deterioration over time, potentially resulting in leaks or accidental radiation exposure. Maintaining a comprehensive inventory of radioactive waste is essential for effective waste management, regulatory compliance, and ensuring overall safety. Such an inventory allows organizations and authorities to monitor and control radioactive waste from its creation to its final disposal. Conducting statistical analysis on the radioactive waste inventory provides valuable insights for waste management planning, regulatory compliance, and decision-making. This analysis can help identify areas needing improvement, develop effective waste management strategies, and ensure the safe and efficient handling of radioactive waste.

1. Introduction

The management of Disused Sealed Radioactive Sources (DSRS) poses a critical challenge in ensuring environmental safety and regulatory compliance. These sources, although retired from active use, still contain radioactive materials that require meticulous handling to mitigate potential risks associated with their storage and disposal. Effective DSRS management is vital to protecting public health, safeguarding the environment, and adhering to stringent regulatory standards.

This study focuses on the comprehensive review and reorganization of the DSRS inventory to enhance the accuracy and reliability of waste management practices in Malaysia. The primary goal is to establish a structured and optimized inventory database that supports robust decision-making for the disposal and management of these sources. Through a detailed

IOP Conf. Series: Materials Science and Engineering **1326** (2025) 012013 doi:10.1088/1757-899X/1326/1/012013

assessment of current records, efforts were made to address data gaps, reduce uncertainties, and align practices with regulatory requirements. This initiative aims to improve the overall efficacy of Malaysia's DSRS management system.

A key component of this paper involves compliance with the Atomic Energy Licensing for Low Activity Radioactive Materials (Exemptions) Order 2020, which specifies exemption limits for low-activity radioactive materials. By identifying materials that fall below these limits, resources can be prioritized for higher-risk waste, allowing for more effective allocation of efforts and the selection of appropriate disposal methods.

The findings from this study are expected to not only optimize existing data but also ensure that disposal practices adhere to international best practices. This will enhance Malaysia's ability to safely and sustain radioactive waste, paving the way for a more efficient approach to handling disused radioactive sources.

2. Literature review

A radioactive source contains radioactive material from specific radionuclides (unstable elements that emit radiation), which vary depending on the intended application for which the source is produced. A sealed radioactive source is an encapsulated radioactive material, generally in the form of a small piece of metal and is considered harmless. IAEA (2024) defines a Disused Sealed Radioactive Source (DSRS) as a source that is no longer in use and has no intention of being used again in the future in a practice authorized by the authorities. If lost or improperly managed, disused sealed sources can pose a threat to human health and the environment. Damage to the source capsule can lead to the release and spread of radioactive material, causing environmental pollution and potential radiological harm to individuals.

Next, DOE (2012) explained that radioactivity is the process by which some unstable atoms spontaneously emit radiation, typically involving alpha or beta particles, often accompanied by gamma rays. This occurs when an atomic nucleus is unstable and needs to release energy to transition to a more stable configuration. When the nucleus emits radiation or disintegrates, the radioactive atom (radionuclide) changes into a different nuclide. This process, known as radioactive decay, continues until the forces within the nucleus reach a balanced state. Every radionuclide emits radiation at its specific rate, which is measured in terms of half-life. Radioactive half-life is the time required for half of the radioactive atoms present to decay. Some radionuclides have half-lives of mere seconds, but others have half-lives of millions of years.

Sealed and unsealed radioactive sources are used for a variety of purposes, and they incorporate a wide range of radionuclides and amounts of radioactive material. Therefore, IAEA Safety Standards created a Safety Guide called Categorization of Radioactive Sources that provides a risk based on ranking of radioactive sources and practices into five categories. The categorization system is based on the concept of dangerous sources which are quantified in terms of D-values. The D value is the radionuclide-specific activity of a source which if not under control, could cause severe deterministic effects for a range of scenarios that include both external exposure from an unshielded source and internal exposure following dispersal of the source material [6]. In this categorization system, Category 1 sources are deemed the most hazardous due to their potential to pose a very high risk to human health if not handled safely and securely. Conversely, Category 5 sources are considered the least hazardous. However, even these sources could exceed the dose limits if not properly managed, therefore they must remain under strict regulatory control [5].

In addition, according to the IAEA (2009), radioactive waste is categorized based on its radioactivity levels, half-lives, and potential hazard. According to AELB (2011), the primary categories are Exempt Waste (EW), which contains minimal radioactivity and poses little risk;

Very Short-Lived Waste (VSLW), which has radionuclides with half-lives up to 100 days and requires temporary storage; Very Low-Level Waste (VLLW), which has low radioactivity and can be disposed of in landfills; Low-Level Waste (LLW), which contains higher radioactivity and requires near-surface disposal; Intermediate-Level Waste (ILW), which necessitates shielding and deeper disposal; and High-Level Waste (HLW), which includes spent nuclear fuel and requires deep geological repositories for safe long-term storage. The IAEA's classification of radioactive waste is crucial for ensuring safety, regulatory compliance, and efficient resource use. It guarantees that each type of waste is managed according to its specific risks, thereby protecting human health and the environment [7].

The AELB (2020), Malaysia's regulatory authority for radiation protection and nuclear safety, establishes standards for managing radioactive waste, including setting exemption levels. These exemption levels are thresholds that determine which radioactive materials or waste are considered so low in activity that they do not pose a significant risk to health or the environment and, therefore, do not require stringent regulatory control. Materials or waste below these exemption levels are generally not subject to rigorous regulatory requirements. This can include certain low-level radioactive materials used in various industries, research, or medical applications. However, even materials below the exemption level must meet specific conditions. They must be handled and disposed of in a way that ensures they do not pose a significant risk to public health or the environment.

3. Methodology

3.1. Categorization of radioactive sources

Disused Sealed Radioactive Sources (DSRS) are classified into five categories by comparing the actual activity (A) to a specific D-value for each radioactive source as shown in Table 1 below. Proper categorization of the sources is crucial for ensuring safety, as it allows for the appropriate management strategies to be applied based on the level of risk and specific characteristics of the waste.

Table 1. Category of radioactive sources based on A/D ratio [4]

Category of Radioactive Sources	Activity Ratio (A/D)
1	$A/D \geq 1000$
2	$1000 \geq A/D > 10$
3	$10 \geq A/D > 1$
4	$1 > A/D \geq 0.01$
5	$0.01 > A/D$

3.2. Classification of radioactive waste

DSRS are waste categorized into different classes based on the types of radionuclide emitters, total activity, and half-life. Table 2 below presents the various classes of radioactive waste along with their descriptions. Classification ensures that waste is handled, stored, and disposed of according to its radioactivity level and associated risks, thus protecting human health and the environment. Different waste categories require specific disposal methods. Classification ensures that each type of waste is disposed of using the appropriate techniques and facilities.

Table 2. Class of radioactive waste and its description [1]

Class of Radioactive Waste	Description
Cleared Waste	Materials containing levels of radionuclides at activity concentrations less than those specified in the Second Schedule.
Low Level (Short Lived)/Decay Waste	Low level radioactive waste containing short lived radionuclides only (half lives less than 100 days) that will decay to clearance levels within three years after the time of its generation.
Low and Intermediate Level Short Lived Waste (LILW-SL)	Radioactive waste which will not decay to clearance levels within three years containing beta/gamma emitting radionuclides with half lives less than thirty years or alpha emitting radionuclides with an activity concentration less than 400Bq/g and a total activity less than 4000Bq in each radioactive waste package.
Low and Intermediate Level Long Lived Waste (LILW-LL)	Radioactive waste containing radionuclides with activity concentrations more than LILW-SL, but which does not generate heat at above 2kW/m^3.
High Level Waste (HLW)	Radioactive waste containing radionuclides with activity concentrations more than LILW-SL, but which generates heat at above 2kW/m^3.

3.3. Exemption limit of radioactive waste

According to AELB (2020) guidelines, DSRS must adhere to specific exemption conditions for low-activity radioactive materials. For materials weighing one tonne or less, the activity concentration per unit weight (Bq/g) of an individual radionuclide listed in column (1) of the First Schedule must not exceed the level specified in column (2). Additionally, the total activity (Bq) of any individual radionuclide present on the premises at any one time must not exceed the level specified in column (3) of the First Schedule. This paper ensures that the total activity of any individual radionuclide on the premises remains below the specified level in column (3). Materials exceeding these exemption limits will undergo additional processing before being safely disposed of into the environment. This precautionary measure is essential to mitigate any potential radiation risks to public health and the environment. For precise details on the exemption limits, including those in columns (1), (2), and (3) of the First Schedule, reference should be made to the AELB (2020) guidelines.

4. Results and discussion

4.1. Analysis of DSRS data in 2019

In 2019, a total of 273 units of radioactive materials DSRS were collected from 53 services throughout the year. Of the total radionuclides collected, 267 units have been identified, while the other 6 units need to be verified by physical observation. Figure 1 below shows the total quantity of DSRS based on different radionuclides collected in 2019. The total current activity for all radionuclides is 191.85 GBq.

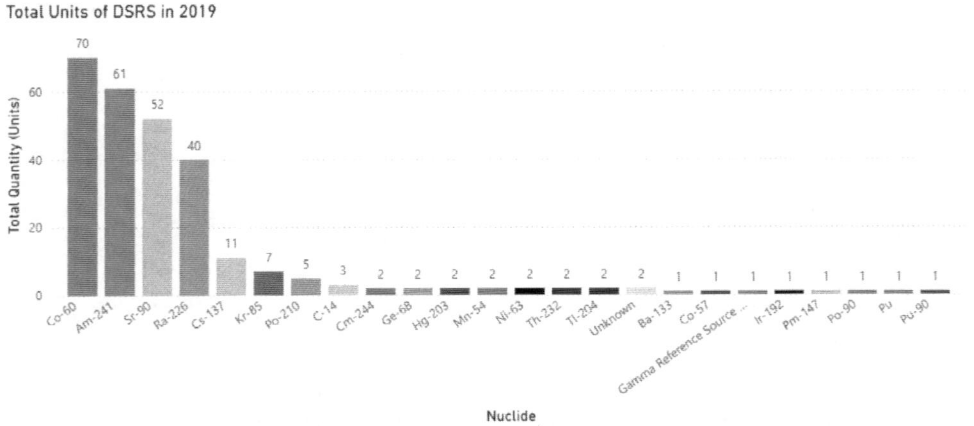

Figure 1. Total units of DSRS in 2019

Figures 2 and 3 below respectively show the total quantity of DSRS that had been used for various applications and sectors collected in 2019. Based on Figure 2, most DSRS were used in experiments, totaling 190 units, followed by gauges with 28 units of DSRS. Additionally, referring to Figure 3 below, a total of 190 units of DSRS have been used in the education sector, followed by 57 units from gauges, 24 units from smoke detectors, and 1 unit each from the medical sector and lightning arresters. Besides, the DSRS had been categorized into 5 categories and as shows in Figure 4 below, 35 units DSRS were categorized in Category 4 and 238 units were categorized in Category 5.

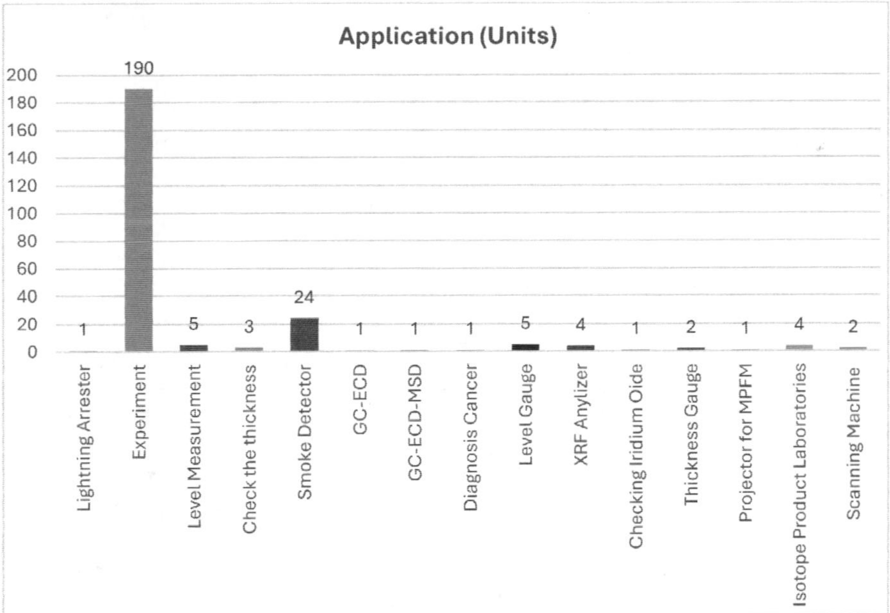

Figure 2. Total quantity of DSRS used for various applications in the year 2019

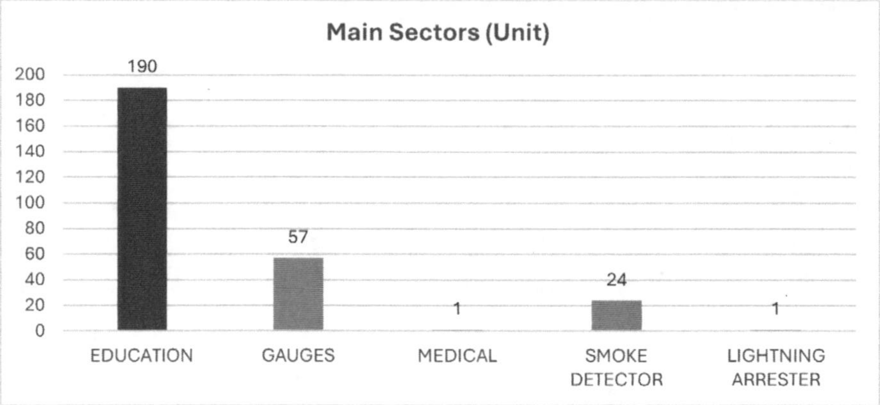

Figure 3. Total quantity of DSRS used in different sectors in the year 2019

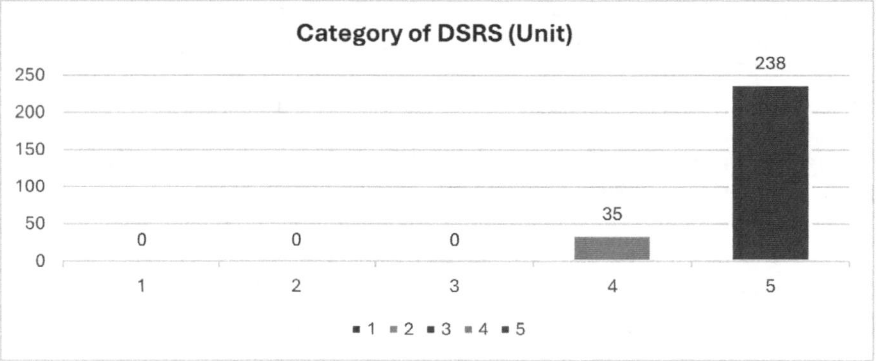

Figure 4. The category of radioactive source DSRS in year 2019

Table 3. Total units and total current activity of radionuclides 2019 after exemption

Radionuclides	Total Units	Total Activity (GBq)
Cs-137	11	44.99836145
Co-60	26	33.98269782
Am-241	60	9.904640435
Sr-90	52	0.004244773
Tl-204	1	7.55224E-05
Kr-85	7	101.7860817
Ni-63	2	0.93450787
Ge-68	1	0.000583537
Th-232	2	7.4E-05
Po-90	1	0
Ra-226	41	0.01090521
Gamma Reference Source Set QCr2	1	0
Unknown	2	0
Total	207	191.6221

Table 3 above is a table that contains radionuclides along with the number of units and the total current activity that need to be managed during the disposal process after undergoing exemption. Based on Table 3, there are 13 types of radionuclides with a total of 207 units and a current total activity of 191.6221 GBq that still need to undergo further processes before they are safe to be released into the environment for DSRS 2019. Figure 5 below illustrates effectively highlights the distribution and significance of different radionuclides in the disposal process post-exemption. The disparities between unit counts and total activities suggest that while some radionuclides are present in large numbers, their impact varies significantly in terms of radioactivity, necessitating tailored approaches for safe management and disposal.

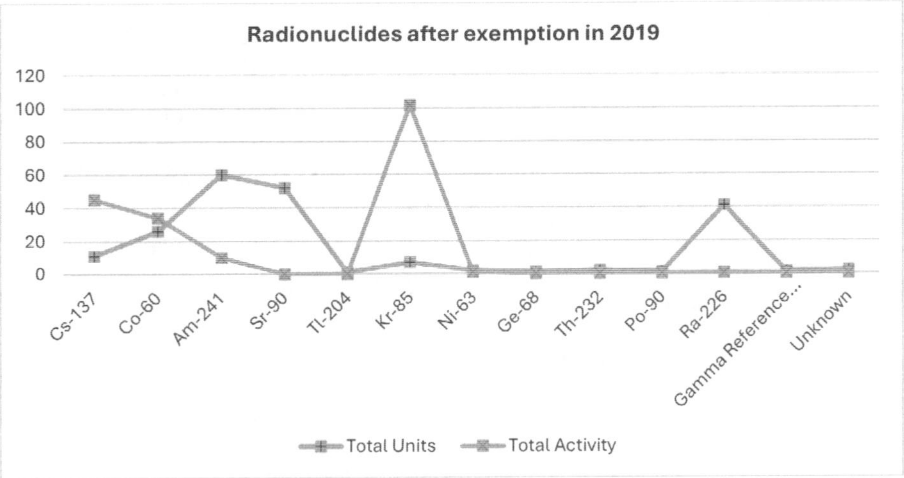

Figure 5. Radionuclides of 2019 after exemption

4.2. Analysis of DSRS data in 2020

In 2020, a total of 110 units of DSRS were collected from 21 services throughout the year. Figure 6 below illustrates the number of DSRS based on different radionuclides that were collected, with a total current activity of 140.18 GBq for all radionuclides.

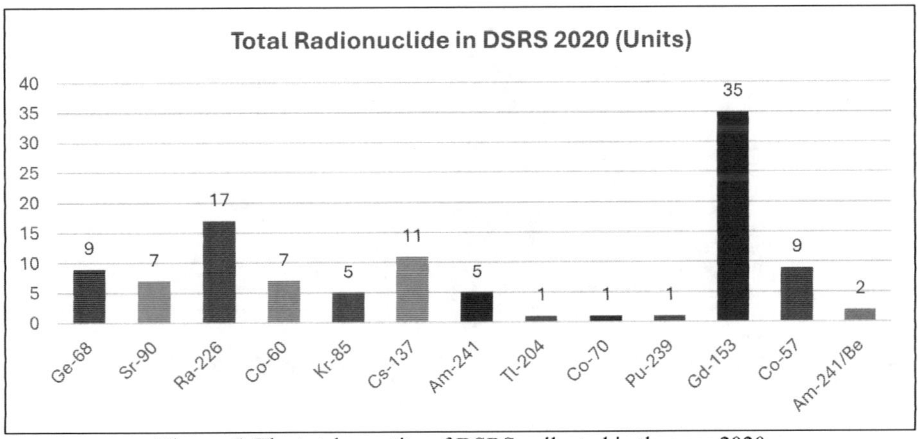

Figure 6. The total quantity of DSRS collected in the year 2020

IOP Conf. Series: Materials Science and Engineering **1326** (2025) 012013 doi:10.1088/1757-899X/1326/1/012013

Figure 7 illustrates the distribution of DSRS across various applications. As indicated in Figure 7 below, the majority of DSRS were used in the medical field, totalling 42 units out of 110, followed by 18 units from AELB confiscations. Notably, the applications with the fewest DSRS radioactive materials used in 2020 were fixed gauge, Iridium Oxide screening, cancer therapy, and sample irradiation, each with only 1 unit.

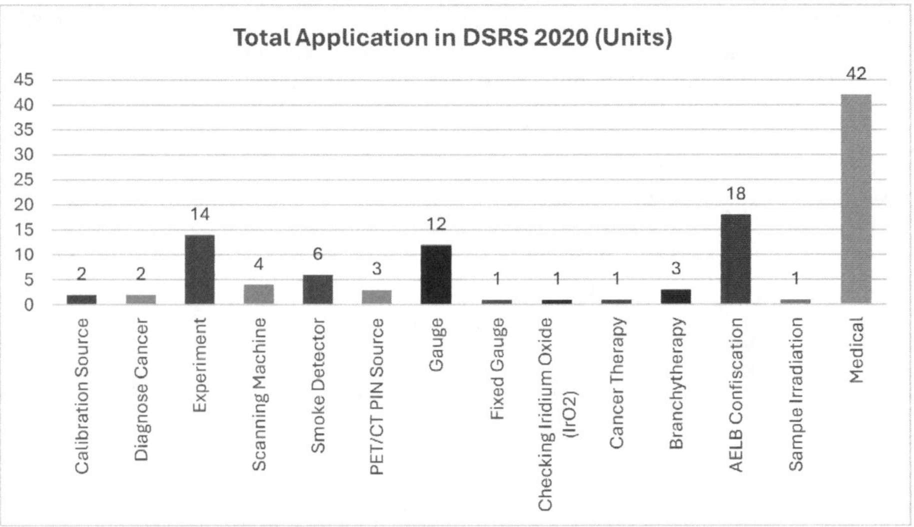

Figure 7. Total quantity of DSRS used from various applications

Based on Figure 8, most DSRS were used in the medical sector which had total of 48 units of DSRS followed by 40 units of DSRS from gauges in industry, 16 units from cup source and 6 units from smoke detectors.

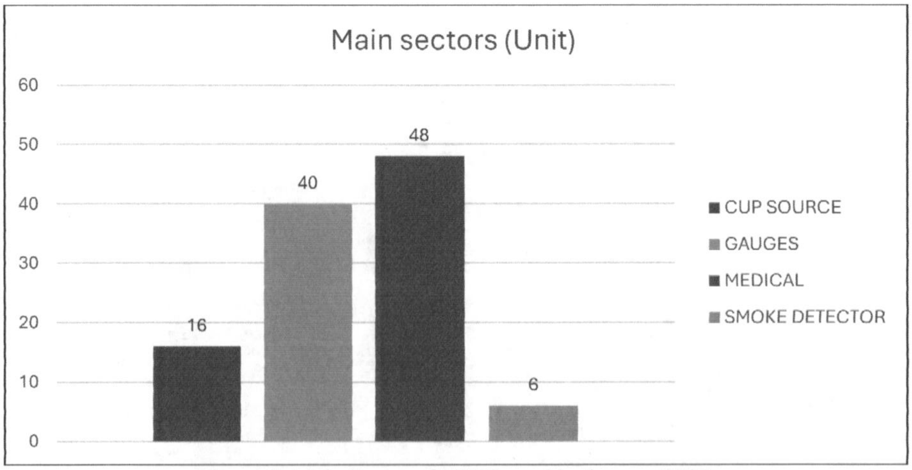

Figure 8. Main Sector of DSRS used in the year 2020

Table 4. Total units and total current activity of radionuclides 2020 after exemption

Radionuclides	Total Units	Total Activity (GBq)
Ni-63	3	1.148100521
Sr-90	35	1.224689997
Cs-137	13	1979.831913
Co-60	23	10892.31807
Am-241	49	12.19508212
Ra-226	52	4.459137308
Ge-68	12	0.055420531
Co-57	1	0.00134424
Tl-204	3	8.39427E-05
Co-70	2	0
Pu-239	6	0.001108481
Total	44	140.0708

Table 4 above provides information on the radionuclides that need to be managed during disposal following their exemption in 2020. It lists 11 types of radionuclides, totaling 44 units with a combined current activity of 140.0708 GBq. These radionuclides require additional processing before they can be safely released into the environment for DSRS 2020.

4.3. Analysis of DSRS data in 2021

In 2021, a total of 247 units of DSRS had been provided and collected from 34 services throughout the year. Figure 9 below illustrates the radionuclides that been managed by collected from various sector in Malaysia and then managed by WasTeC in the year 2021.

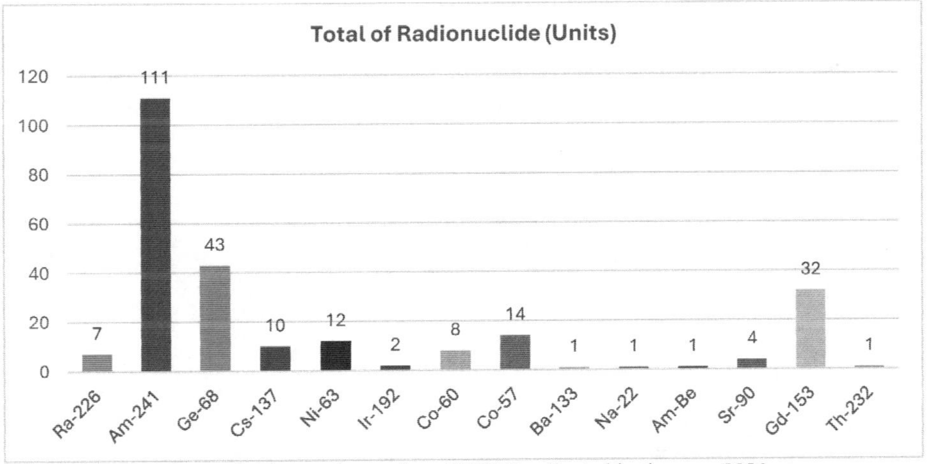

Figure 9. The total quantity of DSRS collected in the year 2021

The total of current activity of all radionuclides is 63.30 GBq. Figure 10 and 11 below show the quantity of DSRS used in different applications and main sectors where the DSRS was mostly used in respectively. Additionally, in 2021, the majority of DSRS were used as smoke detectors, totaling 96 units, followed by 76 units used in the medical field as shown in Figure 10 below.

IOP Conf. Series: Materials Science and Engineering **1326** (2025) 012013 doi:10.1088/1757-899X/1326/1/012013

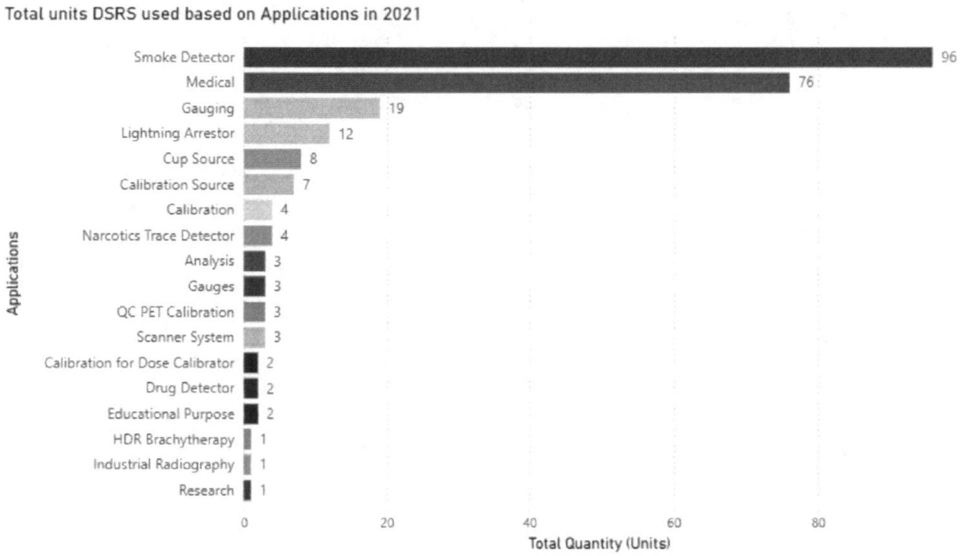

Figure 10. The quantity of DSRS used in various applications in 2021

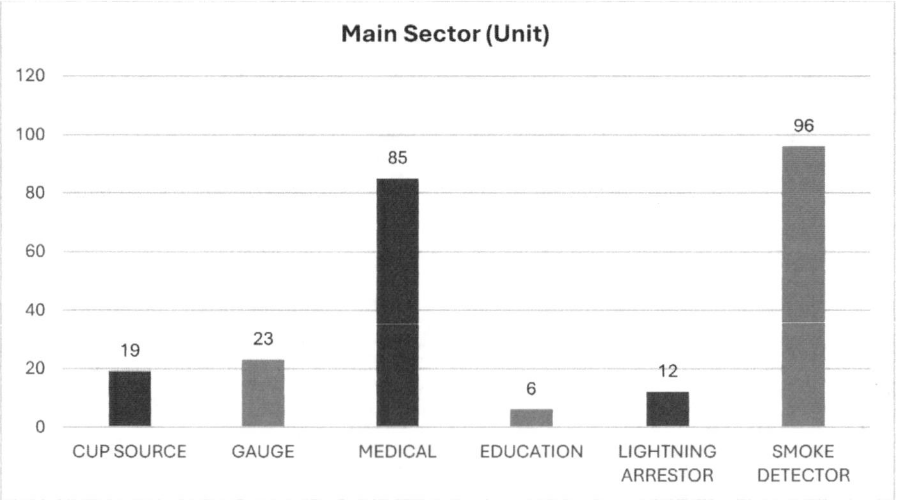

Figure 11. Main Sector of DSRS used in the year 2021

Table 5. Total units and total current activity of radionuclides 2021 after exemption

Radionuclides	Total Units	Total Activity
Ra-226	7	0.055305397
Am-241	111	12.53522617
Ge-68	9	0.003433605
Cs-137	10	9.936015033
Ni-63	12	4.938274005
Ir-192	2	0.66579357
Co-60	5	0.852743774
Co-57	4	0.192576174
Sr-90	4	0.000309034
Ba-133	1	0.002758629
Am-Be	1	34.07547834
Total	166	63.2579

Table 5 above shows the radionuclides, including the number of units and the total current activity, that need to be managed during the disposal process after undergoing exemption in 2021. According to Table 5, there are 11 types of radionuclides, totaling 166 units with a combined current activity of 63.2579 GBq, which still require further processing before they can be safely released into the environment for DSRS 2021.

4.4. Analysis of DSRS data in 2022

In 2022, a total of 343 units of DSRS were collected from 50 different services. The total current activity for this year was 191.13 GBq. As highlighted in Figure 12 below, the cup source sector had a significantly higher usage of DSRS with 173 units, compared to other sectors. The smoke detector sector also had a notable number of units at 76. The gauge, medical, education, and lightning arrestor sectors had comparatively lower usage, with 48, 24, 17, and 5 units respectively.

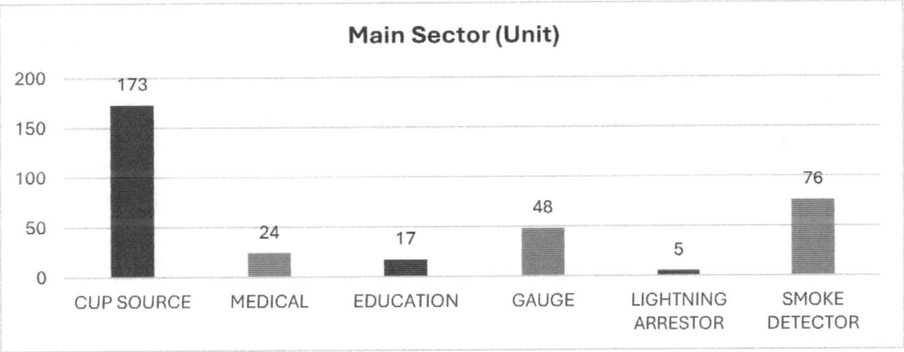

Figure 12. Total number of DSRS used in different sector (unit)

Table 6. Total units and total current activity of radionuclides 2022 after exemption

Radionuclides	Total Units	Total Activity (GBq)
Ge-68	7	0.023652647
Am-241	110	115.8242854
Ra-226	24	2.511322202
Sr-90	26	0.202111603
Ni-63	18	5.901303719
Cs-137	26	16.8109663
Gd-153	4	0.042542383
Pm-147	1	0.770494465
Kr-85	6	42.44844826
Co-57	1	0.005005832
Co-60	4	0.000684523
Pb-210	1	5.55664E-05
Fe-55	4	0.203316529
Pu-238	2	6.206173187
Pu-239	1	0.000184794
Ba-133	1	0.002841641
Cf-252	3	0.053892724
Total	239	191.0073

Table 6 above outlines the radionuclides that require management during disposal after their exemption in 2022. It indicates that there are 17 different types of radionuclides, totalling 239 units with a combined current activity of 190.0073 GBq. These radionuclides must undergo additional processing before they can be safely released into the environment for DSRS 2022.

4.5. Analysis of DSRS data in 2023

In 2023, WasTeC collected and managed a total of 239 units of DSRS from 44 services throughout the year. The total current activity for 2023 was 12,915.56 GBq, a notable increase compared to previous years. This significant rise in activity is primarily due to the inclusion of two Co-60 units used for calibration (teletherapy), which together contributed a high activity of 10,915.77 GBq. As illustrated in Figure 13 below, the education sector predominantly utilized DSRS, accounting for 168 units, a figure substantially higher than other sectors. The lightning arrester sector followed with 31 units, and the cup source sector had only 16 units.

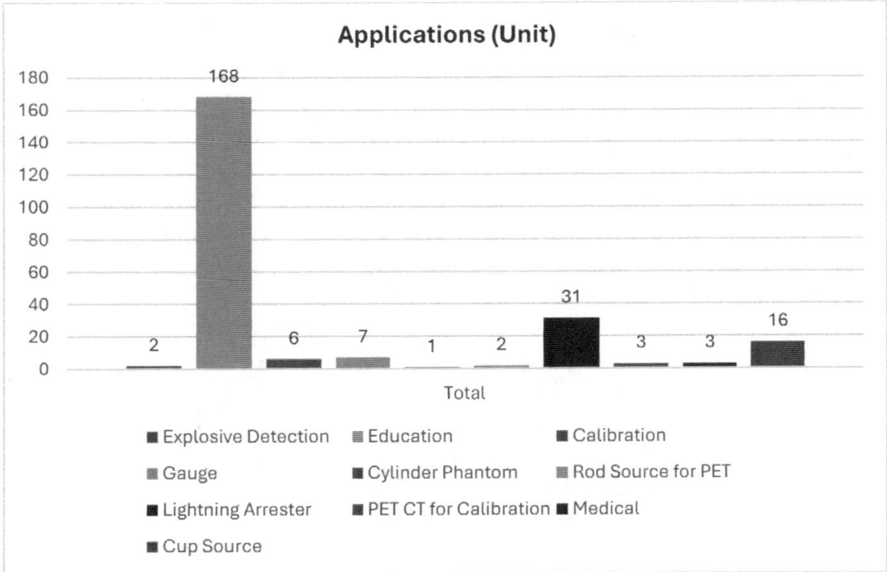

Figure 13. Total units of DSRS used in various applications in the year 2023

Table 7. Total units and total current activity of radionuclides 2023 after exemption

Radionuclides	Total Units	Total Activity
Ni-63	3	1.148100521
Sr-90	35	1.224689997
Cs-137	13	1979.831913
Co-60	23	10892.31807
Am-241	49	12.19508212
Ra-226	52	4.459137308
Ge-68	12	0.055420531
Co-57	1	0.00134424
Tl-204	3	8.39427E-05
Co-70	2	0
Pu-239	6	0.001108481
Total	199	12891.2350

Table 7 above details the radionuclides needing management during disposal following their exemption in 2023. It shows a total of 11 radionuclide types, comprising 199 units with an aggregate current activity of 12891.2350 GBq. These radionuclides must be further processed to ensure they can be safely released into the environment for DSRS 2023.

4.6. *Comparison of DSRS inventory from 2019 to 2023*

From 2019 to 2023, the number of DSRS units managed by WasTeC exhibited notable fluctuations. It began at 273 units in 2019 with a total current activity of 191.85 GBq, decreased to 110 units in 2020 with 140.18 GBq, then rose to 247 units in 2021 with 63.30 GBq. In 2022, the units increased to 343 with a current activity of 191.13 GBq, before dropping to 239 units in 2023, which saw a dramatic surge in total current activity to 12,915.56 GBq. Figures 14 and 15 below illustrate the changes in DSRS units and total current activity over these years.

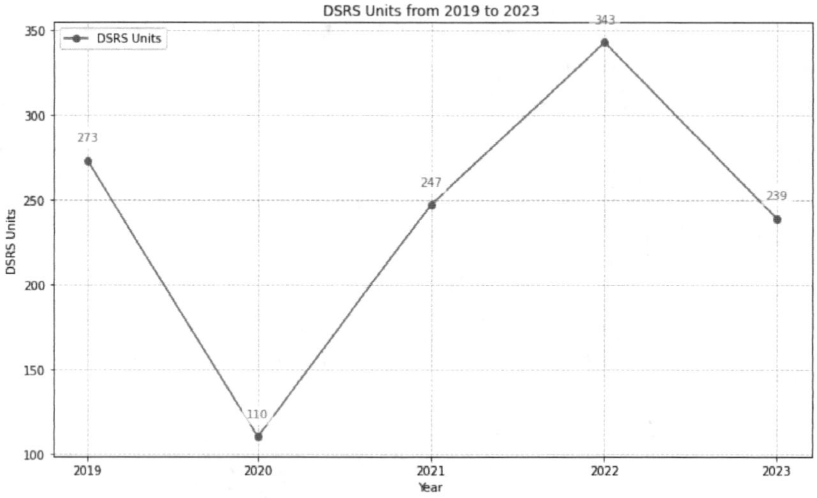

Figure 14. Total units of DSRS from 2019 until 2023

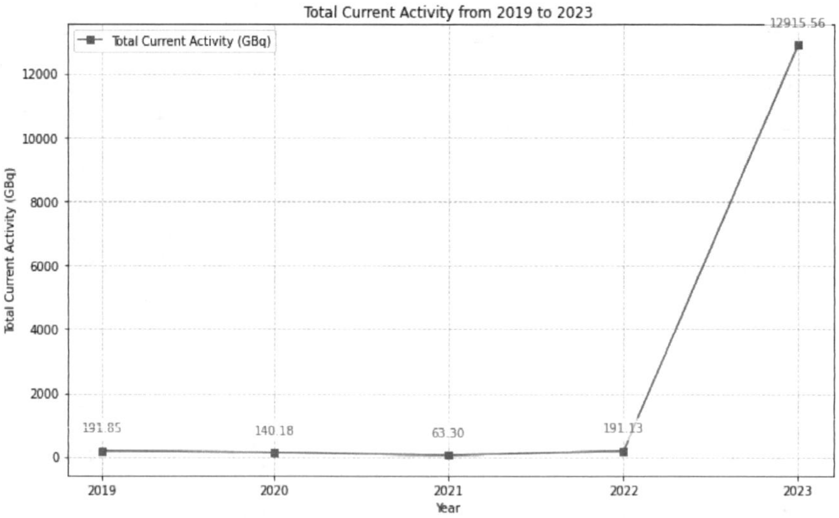

Figure 15. Total current activity (GBq) of DSRS from 2019 until 2023

Table 8. Radionuclides for DSRS 2019 until 2023 after exemption

Year	Radionuclides (after exemption)	Total Units	Total Activity (GBq)
2019	Cs-137, Co-60, Am-241, Sr-90, Tl-204, Kr-85, Ni-63, Ge-68, Th-232, Po-90, Ra-226, Gamma Reference Source Set QCr2, Unknown	207	191.6221723
2020	Sr-90, Ra-226, Co-60, Kr-85, Am-241, Tl-204, Co-70, Pu-239, Am-241/Be, Cs-137, Ge-68	44	140.0708272
2021	Ra-226, Am-241, Ge-68, Cs-137, Ni-63, Ir-192, Co-60, Co-57, Sr-90, Ba-133, Am-Be	166	63.25791373
2022	Ge-68, Am-241, Ra-226, Sr-90, Ni-63, Cs-137, Gd-153, Pm-147, Kr-85, Co-57, Co-60, Pb-210, Fe-55, Pu-238, Pu-239, Ba-133, Cf-252	239	191.0072818
2023	Ni-63, Sr-90, Cs-137, Co-60, Am-241, Ra-226, Ge-68, Co-57, Tl-204, Co-70, Pu-239	199	12891.23495

Table 8 outlines the radionuclides of DSRS from 2019 to 2023 that need to be managed after exemption before safe disposal into the environment. The data reveals notable variations over these years. In 2019, there were 207 units with a total activity of 191.6222 GBq. This number dropped sharply to 44 units in 2020, with a total activity of 140.0708 GBq. The total units then increased to 166 in 2021, although the activity decreased to 63.2579 GBq. By 2022, the number of units reached its peak at 239, with a total activity of 191.0073 GBq. In 2023, there was a significant rise in both the number of units and activity, with 199 units and an activity level of 12891.2349 GBq. This represents a substantial increase compared to previous years, highlighting a dramatic rise in radionuclide activity by 2023.

For the time span of 2019 to 2023, at times the Disused Sealed Radioactive Sources (DSRS) collected in Malaysia indicated a varying trend that was impacted by a number of factors including operational challenges, policy implementations, and post-pandemic recovery efforts. The year 2019 saw the most significant improvement with a total of 273 units, which was a reflection of the predetermined collection level. Nevertheless, a steep fall took place in 2020 with only the sum of 110 units being collected. The dismissal was mostly ascribed to the limitations imposed and the interruptions caused by the Covid-19 pandemic, as it was not the only problem that incurred people from providing the regular course of action of collecting and managing.

With Malaysia's adaption to the new reality, 2021 acted as a year of recovery during which 247 units of DSRS were collected successfully. Thus, the progress was seen as proof of the reestablishment of the production process. The subsequent year revealed visible work on backlogged DSRS with the peak figure of 343 units being achieved in 2022. Surges like these were probably due to the intensified collection campaigns and strategic management improvements. The slump, in 2023, saw only 239 units being collected which was less than the previous year; nevertheless, it was still a large number compared to before the pandemic.

As for the concrete information on the types of DSRS, this paper shows only a small number of details, yet it points out some radionuclides that are usually found in the received sources. These are Cobalt-60 (Co-60) that is widely used in both medical and industrial, Caesium-137 (Cs-137) that is generally used for the calibration of equipment and well as for radiography Iridium-192 (Ir-192) is exclusively used in industrial radiography field. The fact that these radioisotopes keep showing up in the certificates demonstrates the urgency with which these sectors should handle the risk of managing and disposing of the toxic waste properly.

The variable amounts that go up and down and the types that come over and over of radioactive elements that have been collected show that DSRS management is very dynamic in Malaysia. Trends showing this very thing emphasize the fact that there should be strong policies and methods that are aimed at the safe handling, storage, and disposal of radioactive sources.

5. Conclusions

The thorough review and reorganization of the radioactive waste inventory have been pivotal in enhancing the accuracy, reliability, and compliance of our Disused Sealed Radioactive Sources (DSRS) management system. By meticulously verifying, assessing, and correcting inventory records, we have established a robust foundation for effective waste management. The application of best practices in uncertainty management has ensured that the data is both credible and actionable. Besides that, the identification of waste materials falling below the exemption limits outlined in the Atomic Energy Licensing for Low Activity Radioactive Materials (Exemptions) Order 2020 has facilitated a streamlined regulatory approach, enabling us to focus resources on higher-risk materials. This targeted approach, combined with an optimized inventory database, has informed the selection of appropriate disposal methods.

The fluctuating quantities of DSRS collected and the consistent types of radionuclides involved reflect the dynamic nature of DSRS management in Malaysia. These trends emphasize the importance of implementing robust policies and strategies aimed at ensuring the safe handling, storage, and disposal of radioactive sources. The data also underscores the need for adaptable management practices to address collection rate variations and handle backlogs efficiently. As Malaysia continues to enhance its waste management system, a strong focus on regulatory compliance, technological advancements, and capacity-building will be essential to sustainably managing DSRS and safeguarding public and environmental health. This study serves as a stepping stone toward achieving a more comprehensive and sustainable approach to managing radioactive waste in Malaysia.

6. References

[1] AELB 2011 *Atomic Energy Licensing (Radioactive Waste Management) Regulations 2011: Federal Government Gazette* P.U.(A) 274 1 – 64.
[2] AELB 2020 *Atomic Energy Licensing (Low Activity Radioactive Material) (Exemption) Order 2020: Federal Government Gazette* P.U.(A) 256 1 – 65.
[3] DOE 2012 What is radioactivity? *DOE Openness: Human Radiation Experiments: Roadmap to the Project*: *ACHRE Report* https://ehss.energy.gov/ohre/roadmap/achre/intro_9_2.html.
[4] IAEA 2003 Categorization of radioactive sources, *Categorization of Radiation Sources: Revision of IAEA-TECDOC-1911: IAEA-TECDOC-1344* (Vienna, Austria) 1 – 34. https://www-pub.iaea.org/MTCD/Publications/PDF/te_1344_web.pdf.
[5] IAEA 2005 Categorization of radioactive sources *IAEA Safety Standards, Safety Guide No. RS-G-1.9.* (Vienna, Austria) 1 – 70 https://www-pub.iaea.org/MTCD/Publications/PDF/Pub1227_web.pdf.
[6] IAEA 2006 Dangerous quantities of radioactive material (D-values), *IAEA-EPR-D-Values 2006* (Vienna, Austria) 1 – 154 https://www-pub.iaea.org/MTCD/Publications/PDF/EPR_D_web.pdf
[7] IAEA 2009 Classification of radioactive waste, *IAEA Safety Standards, General Safety Guide No. GSG-1.* (Vienna, Austria) 1 – 68 https://www-pub.iaea.org/MTCD/Publications/PDF/Pub1419_web.pdf

[8] IAEA 2024 Disused sources *Nuclear safety, and security: Radioactive waste and spent fuel management* https://www.iaea.org/topics/disused-sources#:~:text=Sealed%20radioactive%20sources%20are%20used,practices%20they%20were%20authorized%20for.

AUTHOR INDEX

Institute of Physics
Dirac House, Temple Back
Bristol BS1 6BE UK

ISSN: 1757-8981
ISBN 979-8-3313-2005-8

CURRAN ASSOCIATES INC.
proceedings.com

16th Symposium of Structural Dynamics (DYNKON 2019)

MATEC Web of Conferences Volume 285 (2019)

Kombornia, Poland
22 - 24 May 2019

edp sciences